NUMERACY ACROSS THE CURRICULUM

NUMERACY ACROSS THE CURRICULUM

Research-based strategies for enhancing teaching and learning

Merrilyn Goos
Vince Geiger
Shelley Dole
Helen Forgasz
Anne Bennison

Routledge
Taylor & Francis Group
LONDON AND NEW YORK

All names of teachers and students have been changed

First published 2019 by Allen & Unwin

Published 2020 by Routledge
2 Park Square, Milton Park, Abingdon, Oxon OX14 4RN
605 Third Avenue, New York, NY 10017

Routledge is an imprint of the Taylor & Francis Group, an informa business

Copyright © Merrilyn Goos, Vince Geiger, Shelley Dole, Helen Forgasz and Anne Bennison 2019

All rights reserved. No part of this book may be reprinted or reproduced or utilised in any form or by any electronic, mechanical, or other means, now known or hereafter invented, including photocopying and recording, or in any information storage or retrieval system, without permission in writing from the publishers.

Notice:
Product or corporate names may be trademarks or registered trademarks, and are used only for identification and explanation without intent to infringe.

 A catalogue record for this book is available from the National Library of Australia

Set in 11.5/18 pt Adobe CaslonPro by Midland Typesetters, Australia

ISBN-13: 9781760297886 (pbk)

CONTENTS

	About the authors	vii
	List of figures, tables and case studies	xi
Chapter 1	Understanding numeracy	1
Chapter 2	Towards numeracy across the curriculum	33
Chapter 3	Numeracy in the 21st century	57
Chapter 4	Numeracy demands	79
Chapter 5	Numeracy opportunities	103
Chapter 6	Planning for numeracy across the curriculum	128
Chapter 7	Whole-school approaches to numeracy	151
Chapter 8	Assessing numeracy learning	177
Chapter 9	Challenges in enhancing numeracy	208
	Acknowledgements	233
	References	235
	Index	249

ABOUT THE AUTHORS

Merrilyn Goos is professor of STEM education at the University of Limerick, in the Republic of Ireland, and honorary professor of education at the University of Queensland, Australia, where she was previously head of the School of Education. She is an internationally recognised mathematics educator whose research is well known for its theoretical innovation and strong focus on classroom practice. Her research interests include students' mathematical thinking, the impact of digital technologies on mathematics learning and teaching, numeracy across the curriculum and the professional learning of mathematics teachers. In 2017, Merrilyn and the Numeracy Across the Curriculum team won the Mathematics Education Research Group of Australasia's Award for Outstanding Contribution to Mathematics Education Research.

Vince Geiger is a professor of mathematics education and STEM program director in the Learning Sciences Institute Australia, Australian Catholic University, Brisbane. Before entering tertiary education, Vince was a secondary teacher of mathematics for 22 years, during which time he served as president of the Australian Association of Mathematics Teachers. Since becoming involved in tertiary education, he has received national awards for both teaching and research. Vince has led a range of national projects related to promoting students' and teachers' capabilities in numeracy across the curriculum, mathematical modelling, the use of digital tools to enhance mathematics learning and effective teacher professional learning.

Shelley Dole is professor and head of the School of Education at the University of the Sunshine Coast, in Queensland, Australia. Her research interests include misconceptions associated with learning mathematics and numeracy; the development of proportional reasoning; mental computation; and teacher professional development. She has led two major research projects on proportional reasoning as fundamental to numeracy and has been involved in other projects associated with numeracy in the early and middle years of schooling, mental computation, number sense and invented algorithms. She is a senior fellow of the Higher Education Academy, United Kingdom.

Helen Forgasz is a professor emerita in the Faculty of Education, Monash University, in Melbourne, Australia. Her fields of research expertise include equity and mathematics education, with a particular focus on gender issues; numeracy across the curriculum; and student grouping practices and settings for mathematics learning. Helen has won research awards and several prestigious research grants. Her research findings have been published widely in scholarly and professional journals and books and in the popular media.

Anne Bennison is a lecturer in the School of Education at the University of the Sunshine Coast, in Queensland, Australia. In her doctoral studies, she employed a sociocultural approach to identify ways to support teachers to embed numeracy across the curriculum. Anne received a Dean's Award for Outstanding Research Higher Degree Theses, and articles from this work have been published in major education research journals. Her research interests centre on the professional learning of both pre-service and in-service teachers, with a focus on how teachers translate their professional learning experiences into classroom practice, particularly when addressing numeracy across the curriculum.

LIST OF FIGURES, TABLES AND CASE STUDIES

Figures

1.1	Plan view of a 15-centimetre-square cake tin and a 30-centimetre-square cake tin	8
1.2	Excerpt from a spreadsheet for evaluating tenders	13
1.3	Cartograms of the earth's countries by land area and population, 2018	16
1.4	Cartograms of the earth's countries by gross domestic product wealth, 2018, and absolute poverty, 2016	17
2.1	Three kinds of 'know-how' for being numerate	45
3.1	The 21st Century Numeracy Model	60
3.2	Correlation between selected countries' annual per capita chocolate consumption and the number of Nobel laureates per 10 million population	69
5.1	Excerpt from the 'Understanding movement' sub-strand of the Australian Curriculum: Health and Physical Education for Years 7 and 8	108
5.2	Excerpt from the 'Creating texts' sub-strand of the Australian Curriculum: English for Year 7	110
5.3	Excerpt from the 'Spanish conquest of the Americas' depth study of the Australian Curriculum: History for Year 8	113
5.4	Map of South America showing the location of the Incan empire	115

5.5	Template for analysing numeracy tasks	124
6.1	Scale relating emotions to the pace of reading a poem	148
7.1	Sample numeracy investigation	162
7.2	One teacher's trajectory through the 21st Century Numeracy Model	164
8.1	NAPLAN numeracy test sample item	181
8.2	Australian students' performance on PISA mathematical literacy over time	185
8.3	Question from 2012 released PISA items	187
8.4	Sample LANTITE item	190
8.5	PIAAC numeracy performance by country, 2013–16	191
8.6	Florence Nightingale: the lady of the lamp	194
8.7	Florence Nightingale's polar-area diagram (polar graph)	195
8.8	Balloon filled with one tonne of CO_2	196
8.9	Faces for assessing numeracy dispositions	201
9.1	A framework for identity as an embedder-of-numeracy	219
9.2	'Numeracy for learners and teachers' survey participants' pre- and post-unit confidence levels, 2015–17	229

Tables

1.1	Comparison of steps in tendering a category and young workers' approaches	11
1.2	Numeracy self-assessment survey for all teachers of mathematics	26
1.3	Numeracy self-assessment survey for teachers of disciplines other than mathematics	29
2.1	Questions asked by a person acting numerately across knowledge domains	47

2.2	Australian Professional Standards for Teachers	54
3.1	Dimensions of the 21st Century Numeracy Model	59
3.2	Comparison of prices of Max and Red Rock beer by volume	62
4.1	Examples of numeracy demands in Years 7 and 8 HPE	83
4.2	Mathematical knowledge demands within strands of the arts learning area of the SACSA Framework	89
4.3	Mathematical knowledge demands within learning areas of the SACSA Framework (excluding mathematics)	93
5.1	Examples of how numeracy can be embedded across the curriculum	107
5.2	Population data for the Inca, Spanish and African peoples in South America, 1491–1600	117
5.3	Selected events during the Spanish conquest of the Americas	120
5.4	Significant events and developments during the Spanish conquest of the Americas	121
7.1	Professional development plan for whole-school numeracy	161
7.2	Numeracy audit approach	167
7.3	Survey of teacher perceptions of numeracy	171
7.4	Use of a numeracy filter to plan for embedding numeracy into the history curriculum	172
7.5	Plan for initial development of whole-school numeracy approach	174
9.1	Results of 'Numeracy for learners and teachers' pre- and post-surveys	228

Case studies

1.1	Technology and workplace numeracy	10
3.1	Australia's unemployment	71
3.2	Reducing the size of chocolate bars	73
5.1	Using a pedometer	109
5.2	Writing an opinion piece	111
5.3	Understanding the Spanish conquest of the Americas	114
5.4	Features of timelines	120
5.5	Constructing timelines	121
6.1	Looking, noticing and seeing	132
6.2	Structuring and fit to circumstance	137
6.3	Pedagogical architecture	141
6.4	A numeracy task to support learning in English	144
7.1	Whole-school numeracy in primary and secondary schools	168
8.1	A history, health, mathematics or statistics task	194
8.2	A sustainability or environmental studies (science or geography) task	196

1

Understanding numeracy

> Quantitatively literate citizens need to know more than formulas and equations. They need a predisposition to look at the world through mathematical eyes, to see the benefits (and risks) of thinking quantitatively about commonplace issues, and to approach complex problems with confidence in the value of careful reasoning. Quantitative literacy empowers people by giving them tools to think for themselves, to ask intelligent questions of experts, and to confront authority confidently. These are skills required to thrive in the modern world. (Quantitative Literacy Design Team 2001, p. 2)

Numeracy—sometimes referred to as *quantitative literacy*—is often interpreted narrowly as involving facility with numbers and computation performed with paper and pencil or 'in the head', and many people would regard reliance on an electronic calculator as evidence of innumeracy arising from lack of basic number skills. However, this kind of 'basic skills' definition of numeracy is clearly outdated in the data-drenched, technology-rich world of the 21st century. The constant evolution of knowledge, social structures, work practices and new technologies requires a corresponding evolution—or even revolution—in the ways in which we think about numeracy.

This chapter encourages prospective and practising teachers to identify and refine their personal conception of numeracy. We begin by exploring the origins of numeracy as a concept and comparing some of the common definitions and terminologies for describing numeracy. Next, we look for examples of numeracy in a range of real-life contexts to highlight the important distinction between numeracy and mathematics. We then consider the impact of poor numeracy on young people's life chances to argue that numeracy education must be embedded in *all subjects* across the primary and secondary school curriculums. Finally, we find out how well prepared and confident teachers feel in taking on this task.

WHAT IS NUMERACY?

The idea of numeracy is a relatively recent one. The term was first introduced in the United Kingdom by the Crowther Report (Ministry of Education 1959) and was defined as the mirror image of literacy, but involving quantitative thinking. Another early definition proposed by the UK Cockcroft (1982) Report described 'being numerate' as possessing an at-homeness with numbers and an ability to use mathematical skills to cope confidently with the practical demands of everyday life.

> **Review and reflect 1.1**
>
> 1 Before reading any further, write down your responses to the following prompts to capture your own ideas about numeracy.
>   - Numeracy involves...
>   - A numerate person can...
>   - A numerate person knows...
>   - A numerate person is...
> 2 Compare your responses with those of a colleague and summarise any similarities or differences.

Teachers' conceptions of numeracy

We have carried out the 'What is numeracy?' task presented in Review and reflect 1.1 many times with primary school teachers, teachers of secondary school mathematics and teachers of other secondary school subjects. For the first group of teachers we worked with (see Goos, Geiger & Dole 2011), the most frequent responses to the prompt 'Numeracy involves...' showed an appreciation of the role of *context*, with responses referring to 'everyday connections' and 'application of mathematical processes in everyday practical situations'. These teachers thought that numeracy also involved *problem-solving*—for example, 'solving problems in life' or 'having a repertoire of strategies'. When describing what 'a numerate person can' do, the teachers typically wrote about *mathematical skills* such as 'using numbers to solve problems'. But some also thought that a numerate person can 'sort out how to transfer mathematical knowledge into real life situations' and 'use problem solving skills to help them better understand some aspects of numeracy'. In deciding what 'a numerate person knows', the teachers again commonly alluded to *mathematical knowledge* in the form of specific skills (e.g., 'how to convert currency'), but some emphasised knowing mathematics that was appropriate to a particular task ('how and when to use what skill') or how to use mathematics in everyday life ('understand the odds of Melbourne Cup horses'). Their responses to the prompt 'A numerate person is . . .' again revealed an appreciation of contexts for numeracy (e.g., 'someone who can use numeracy in everyday situations') as well as an emphasis on positive *dispositions* (e.g., 'flexible in their mathematical thinking and confident to take learning risks'). Taken together, their responses suggested that this group of teachers understood numeracy as involving the confident application of mathematical knowledge, skills and problem-solving strategies across a range of everyday contexts.

Numeracy terminology and definitions

Although numeracy is a term used in many English-speaking countries, such as the United Kingdom, Ireland, Canada, South Africa, Australia and New Zealand, in the United States and elsewhere it is more common to speak of *quantitative literacy* or *mathematical literacy*. The Quantitative Literacy Design Team (2001) was formed by the US National Council on Education and the Disciplines to inquire into the meaning of numeracy in contemporary society. This team described quantitative literacy as 'the capacity to deal effectively with the quantitative aspects of life' (p. 6) and proposed that its elements included confidence with mathematics, appreciation of the nature and history of mathematics and its significance

> **Review and reflect 1.2**
>
> Numeracy is a term used mainly in English-speaking countries. Other countries in the world do not have a comparable term for numeracy or have simply adopted the PISA definition of mathematical literacy. Geiger, Goos and Forgasz's (2015) article presents a synthesis of international literature on the concept of numeracy.
>
> 1. Read the article with a partner, paying particular attention to the sections on 'Understandings of numeracy' and 'Facets of numeracy'.
> 2. Working individually, summarise the different ways in which numeracy is understood around the world.
> 3. Compare your summary with your partner's and with your responses to the Review and reflect 1.1 task.

for understanding issues in the public realm, logical thinking and decision-making, use of mathematics to solve practical everyday problems in different contexts, number sense and symbol sense, reasoning with data and the ability to draw on a range of prerequisite mathematical knowledge and tools. The Organisation for Economic Co-operation and Development's (OECD) Programme for International Student Assessment (PISA) offered a similarly expansive definition of mathematical literacy as 'an individual's capacity to formulate, employ, and interpret mathematics in a variety of contexts. It includes reasoning mathematically and using mathematical concepts, procedures, facts, and tools to describe, explain, and predict phenomena. It assists individuals to recognise the role that mathematics plays in the world and to make the well-founded judgements and decisions needed by constructive, engaged and reflective citizens' (OECD 2016, p. 65).

In Australia, educators and policy-makers have embraced a broad interpretation of numeracy similar to the OECD's definition of mathematical literacy. The definition adopted by a 1997 national numeracy conference—'To be numerate is to use mathematics effectively to meet the general demands of life at home, in paid work, and for participation in community and civic life' (DEETYA 1997, p. 15)—became widely accepted in Australia and formed the basis for much numeracy-related research and curriculum development.

FINDING NUMERACY

The Australian description of numeracy highlights its role in helping people meet the general demands of life in three different contexts: home, work, and community and civic life. Teachers who are effective at developing their students' numeracy capabilities are also adept at recognising the numeracy demands of these everyday situations—they have cultivated the ability to *see* numeracy by asking questions about the world around them. What it means to see numeracy in this way is explored in the following examples.

At home

Many examples of numeracy in action can be found in the home. For example, cooking involves measurement of quantities and time (see Review and reflect 1.3); shopping requires the ability to compare prices and estimate value for money; playing team sports means keeping score and predicting the effect of different outcomes on an overall league table.

Review and reflect 1.3

A regular feature in a local newspaper invited readers to write to a well-known chef with questions about recipes. This reader's question, and the reply, caught our eye:

> Q. I am planning to make a small Christmas cake in a six-inch tin (15 cm) and would like to know how to calculate the quantities of ingredients needed.
>
> A. Just break down the recipe accordingly; for example, if your cake recipe is for a 12-inch tin (30 cm), then halve the recipe.

1   Think about the advice given by the cookery expert.
    - Was it good advice?
    - What would you have said?
2   Compare your responses with a partner and discuss the numeracy demands of this task.

To make sense of the newspaper reader's question, we might first ask, What does a 15-centimetre-square cake tin look like? Teachers who engage with this task usually envisage a square tin with a side length of 15 centimetres, but some instead assume the tin is circular and the 15-centimetre measurement refers to the diameter. If we have a 15-centimetre-square tin but a recipe for a 30-centimetre-square tin, the comparison between the tins can be represented as in Figure 1.1, which shows a plan view of the two cake tins looking down from above on the bases of the tins.

In comparing only the bases of the cake tins, we are assuming that the tins are the same depth and that we would like our cake to be the same height whether cooked in the smaller or larger tin. Without doing any calculations

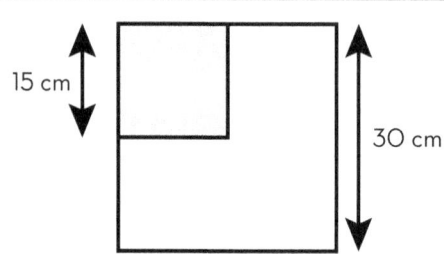

Figure 1.1 Plan view of a 15-centimetre-square cake tin and a 30-centimetre-square cake tin

at all, it is easy to see from Figure 1.1 that the recipe for the 15-centimetre-square tin should be one-quarter, not one-half, of the recipe for the 30-centimetre-square tin. Would we arrive at the same conclusion if the tins were round rather than square? Should we also adjust the baking time in the oven for the smaller cake? Understanding this situation requires a blend of mathematical know-how (but not necessarily knowledge of any formulae), an appreciation of the real-life context of baking and a critical orientation towards what one reads in the newspaper.

At work

All jobs require numeracy. However, students might believe that people in some occupations—such as cleaners, gardeners or factory workers—don't need to be numerate in their jobs, perhaps because they assume that these kinds of work involve no mathematical calculations. But is this really the case?

Mathematics education researchers have investigated the numeracies involved in many different occupational sites, such as building (Bessot 1996), hotels (Kanes 1996), nursing (Coben

> **Review and reflect 1.4**
>
> A gardener wishes to use a 10-litre knapsack sprayer to kill some weeds. The weedkiller instructions say, 'Rate: Apply 15 millilitres per 4 litres of water per 10 square metres'.
>
> 1. Work with a partner to write a clear set of instructions on how to make up 10 litres of diluted herbicide according to the instructions. What area of lawn or garden can be sprayed with this solution?
> 2. Compare your instructions with those written by other colleagues and discuss the numeracy demands of this task.

& Weeks 2014), railway engineering (Wake 2014) and retailing (Jorgensen Zevenbergen 2011). This research shows that numeracy practices are specific to each work context and usually require intuition, estimation, problem-solving and using rules of thumb or tools tailored to specific circumstances (Noss et al. 2000). These are not methods traditionally taught in school mathematics. In fact, it is not the responsibility of school teachers to identify the mathematics needed in different workplaces and to teach this mathematics to their students. Instead, school mathematics needs to build students' understanding of concepts, confidence and adaptive thinking so they can apply their knowledge in a variety of contexts.

The numeracy demands of the workplace are also changing as digital technologies take over many tasks that were previously done manually. Jorgensen Zevenbergen (2011) noted that young people are sometimes criticised for lacking the numeracy understandings and dispositions valued by older workers, such as the ability to carry out accurate mental or written computations. She argued that these older practices might be less relevant in the modern world, because easy access to technologies like calculators, spreadsheets and the internet has led to dramatic changes to curriculum, teaching and assessment in schools and also to the ways in which people live and work. She cited case studies of how young people working in retailing used problem-solving approaches and often deferred the cognitive labour to technology when undertaking tasks. However, this approach often surprised their supervisors, who continued to value traditional practices and tools. Her research found that these older workers were more likely to stress the importance of accurate numerical calculation done by hand or in the head, whereas young workers did not see their jobs in this way—to them, calculation was a menial task that can be delegated to technology, allowing them to develop a stronger orientation towards problem-solving and estimation.

Case study 1.1 Technology and workplace numeracy

The supply manager of a large food-manufacturing company is responsible for setting up contracts with suppliers of ingredients, packaging and services such as electricity, usually after putting this supply work out to tender. He is training two young people in this work with the aim of each of them eventually becoming a category manager responsible for negotiating contracts.

Recently, the supply manager decided to test the young workers by giving them each a category to tender on their own for the first time. The tenders were for contracts to supply plastic wrapping film and machine tape (similar to duct tape or packing tape). He was surprised to see how they transformed the standard steps of tendering a category by using different forms of digital technologies. Table 1.1 summarises the steps and the approaches used by the two young workers.

Table 1.1 Comparison of steps in tendering a category and young workers' approaches

Steps	Young workers' approaches
1 Analyse the spend	Used SAP software and produced a detailed spreadsheet
2 Determine future business need by consulting internal stakeholders	Telephoned to drill down to on-the-floor stakeholders
3 Develop tender documentation	Used company template
4 Identify businesses to invite to tender	Googled using sophisticated search strategies
5 Evaluate tenders	Developed sophisticated spreadsheets with linked worksheets and formulae

Analysing the spend involves identifying what the food-manufacturing company is purchasing now and from whom, and how much is currently being spent. The supply manager usually uses the SAP enterprise resource system, a multifunction software package that manages company finances, purchasing, production planning and stock control. His young workers did so too, but they additionally transferred the information they extracted from the system to a spreadsheet that analysed the spend by quantity, by factory and by supplier.

Determining future business need is usually done by telephoning stakeholders, and this is what the young workers did. Developing tender documentation involves using the company's standard template, a process that the young workers also followed.

The supply manager relies on his previous contacts to identify businesses to invite to tender. However, the young workers—who lacked this experience—instead used sophisticated Google search strategies and identified a much larger group of potential suppliers. They then checked the potential suppliers' websites, extracted the information they needed and conducted telephone interviews with a shortlisted set of companies.

The process of evaluating tenders is usually based on price, with some informal and unquantified allowances made for other criteria. The young workers instead created a comprehensive spreadsheet featuring formulae and both qualitative and quantitative evaluation criteria within linked worksheets (see Figure 1.2).

Questions

Read the accounts of young workers' retail practices in Jorgensen Zevenbergen's (2011) article and use her ideas to interpret the supply management situation outlined above.

1. What evidence is there that the young supply management workers approached the task using numeracy practices (e.g., estimation, problem-solving, use of technology) different from those expected by older generations?
2. What 'old' numeracy practices remained important?

UNDERSTANDING NUMERACY   13

Pricing Schedule										
Product	SAP Material Code	Current Supplier	Current Price	Current UOM	Supplier 1:			Supplier 2:		
					Price	UOM	Weighting	Price	UOM	Weighting
450M X 3000M	517880			Roll						
STRETCHWRAP (450MM X 3600M	503731			Roll						
STRETCHWRAP (450MM X 450M	514664			Roll 2.24kg						
500MM X 1920M X 17UM	516830			Roll						
STRETCHWRAP (500MM X 1420M X 23UM) CAST (MACHINE)	516050			Roll						
STRETCHWRAP (500MM X 500M) (HAND)	513402			Roll 2.3kg						
LABELS										
THERMAL LABEL 125MM X 120MM	514791			1000 labels	$15.96	1000 labels	0.15	$23.75	1000 labels	-9.54
PALLET LABELS YELLOW 100MM X 100M	423177			1000 labels	$19.90	1000 labels	10.00	$41.15	1000 labels	-0.68
WHITE BUTT-CUTT LABELS				1000 labels	$23.39	1000 labels	8.27	$29.25	1000 labels	5.33
HOLD STICKERS				1000 labels	$12.61	1000 labels	10.00			
FLURO LABELS 100 X 100MM				1000 labels	$25.85	1000 labels	10.00	$41.15	1000 labels	4.08
LABEL PERFORATED THERMAL PRINTER 103 X 150 X 1000				1000 labels	$17.68	1000 labels	9.64	$26.88	1000 labels	4.24
							48.06			3.44
TAPE										
MASKING TAPE 50M X 36M		N/A	N/A							
TAPE 'DUMPED'		N/A	N/A							
TAPE 'EMPTY'		N/A	N/A							
TAPE 'HOLD' (BALCK ON YELLOW)		N/A	N/A							
TAPE 'REJECT' (BLACK ON RED)		N/A	N/A							
TAPE 'QUARANTINE' (RED ON YELLOW)		N/A	N/A							
CLEAR TAPE 75M X 38MM GENERAL PURPOSE		N/A	N/A							

Figure 1.2 Excerpt from a spreadsheet for evaluating tenders

In community and civic life

Numeracy is vital for critical citizenship, because almost every public issue depends on data for constructing arguments to inform, persuade or shape decision-making. The news, advertising and entertainment media are rich sources of opportunities to see numeracy in community and civic life and to question the ways in which issues are represented or argued. For example, telephone or online surveys and opinion polls are used to gauge the public's views on issues of interest to political parties or businesses selling goods and services. Uncritical citizens might accept a survey's numerical findings without questioning its design and sampling strategy. The wording and sequencing of survey questions can unintentionally—or intentionally, as in the example in Review and reflect 1.5—lead respondents to a particular answer.

Review and reflect 1.5

The ministerial broadcast

In his first television broadcast Prime Minister Jim Hacker intends to announce the reintroduction of conscription. He says this will be a vote-winner because a Party poll has shown that 64% of the population are in favour. Cabinet Secretary Sir Humphrey Appleby advises Principal Private Secretary Bernard Woolley to issue another poll to show the majority of the population is against reintroducing conscription. Bernard wonders how this can be done and Sir Humphrey explains. (Yes [Prime] Minister Files 2006)

1. Visit the Yes (Prime) Minister Files (2006) page and scroll down to the 'Quotes' section to see the full transcript of the exchange between Sir Humphrey and Bernard Woolley.
2. Visit the website of a polling company, such as Galaxy Research (<www.galaxyresearch.com.au>), OmniPoll (<www.omnipoll.com.au>) or Roy Morgan (<www.roymorgan.com>).
3. Select a polling report for an issue of interest (e.g., voting intentions, relative importance of policy issues, rating of Australia as a place to live, health, finances, relationships, euthanasia, same-sex marriage) and write a few sentences to summarise the main findings.
4. Design a sequence of questions aiming to lead survey respondents towards the opposite view.
5. Try out your questions with a partner to test their effectiveness.

Numeracy is often thought to be about numbers, but numerate citizens also need to understand verbal arguments (as in the example in Review and reflect 1.5) and visual representations of quantitative or spatial information such as graphs, diagrams and maps. Most people would assume that maps represent reality, accurately portraying features of the three-dimensional earth such as area, distance, direction and shape. However, all flat maps distort one or more of these aspects in their representation of the earth. Therefore, the mapmaker must decide the characteristic that is to be shown accurately at the expense of the others or create a compromise among several properties. For example, the map with which most people are familiar, the Mercator projection, distorts area and shape the further one moves towards the North and South poles. This greatly enlarges the size of Europe and shrinks the size of Africa. From looking at this kind of map, people may not realise that what is commonly referred to as Europe has an area less than 20 per cent of the area of Africa. Created in 1569, the Mercator projection preserves compass direction, thus making it very helpful to navigators who sailed from Europe in the 16th century. (See Wilkins & Hicks 2001 for more information on the numeracy of map projections.)

A cartogram is an unusual and powerful kind of map in which area is deliberately not preserved. Instead, the geometry of the map is distorted to convey information about another variable, such as population (see <www.worldmapper.org>). Figure 1.3 shows cartograms depicting world land area and population: in Figure 1.3(b), the areas of countries have been scaled in proportion to their population. For example, Australia's land area is 21 times that of Japan's, but Japan's population is more than 6 times the population of Australia.

(a)

(b)

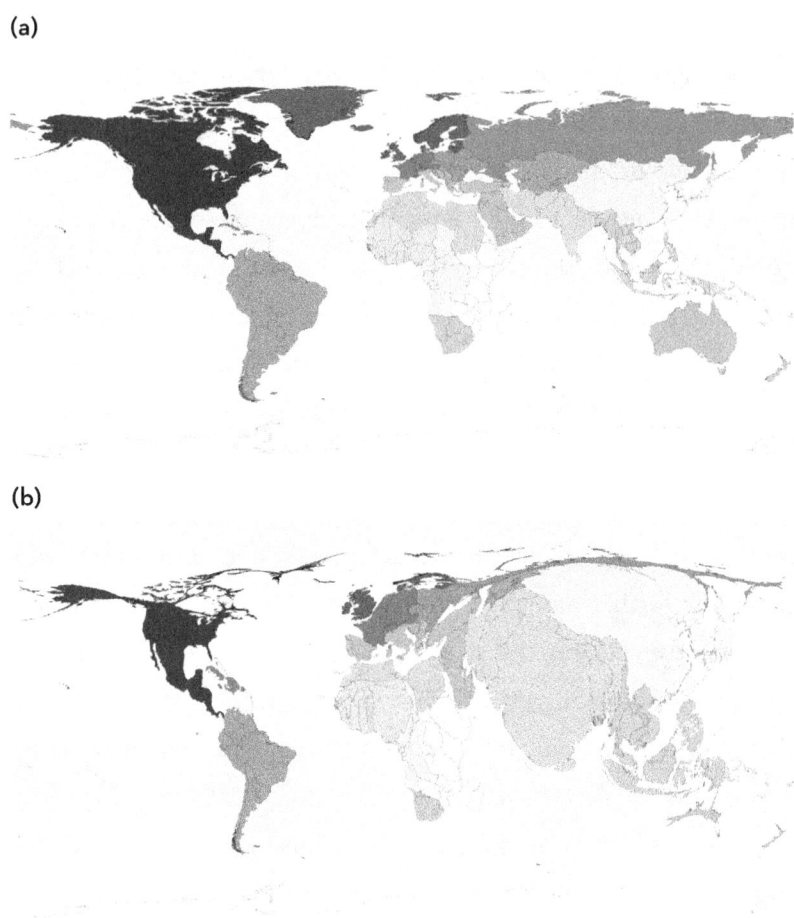

Figure 1.3 Cartograms of the earth's countries by land area (a) and population (b), 2018

Source: Worldmapper.org.

Many other comparisons are possible using different cartograms. Figure 1.4 compares wealth and absolute poverty. Here, wealth is measured as purchasing power, or what can be bought in the country in which the money is earned. Figure 1.4(a) shows that nearly half of the world's wealth adjusted for purchasing power is in North America

(a)

(b)

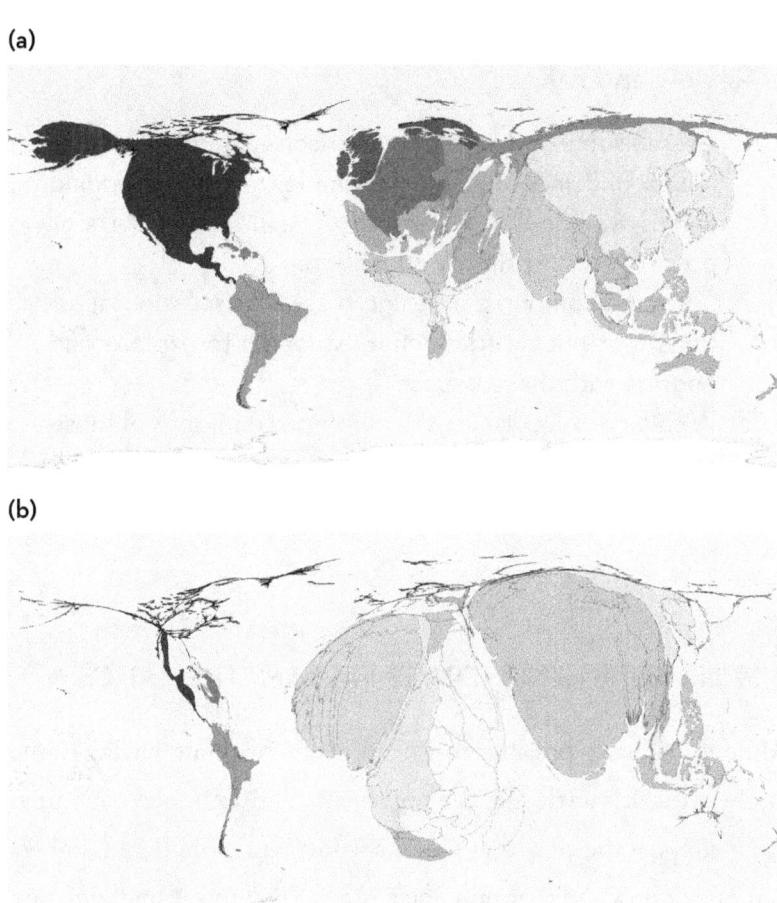

Figure 1.4 Cartograms of the earth's countries by gross domestic product wealth, 2018 (a), and absolute poverty, 2016 (b)

Source: Worldmapper.org.

and Western Europe. Even though many countries in Africa have lower prices, the people living there have very low purchasing power. Figure 1.4(b) shows that the highest proportions of people living in absolute poverty, defined by the World Bank as living on the equivalent of US$1.90 a day or less, are in India and parts of Africa.

> **Review and reflect 1.6**
>
> 1 Create some cartogram comparisons to highlight social issues and inequities—for example, military spending versus war deaths, toy exports versus toy imports or poverty versus primary education.
> 2 For each comparison design a set of discussion questions that encourages school students to explore and engage with these issues.
> 3 With a partner, discuss the numeracy demands of these tasks.

DISTINGUISHING NUMERACY FROM MATHEMATICS

Although it isn't possible to be numerate without having some mathematical knowledge, the arguments and examples presented so far support the view that numeracy involves more than numbers and calculation and that numeracy is not the same as mathematics. Nor should numeracy be considered a 'watered down' version of mathematics. The difference between numeracy and mathematics is rather difficult to capture in words. If you search for definitions of mathematics, you will typically find it described as an abstract science of quantity, structures, space and change. Mathematicians look for patterns and formulate conjectures. They use mathematical proof to resolve the truth or falsity of these conjectures. The work of professional mathematicians can seem abstract and inwards, shutting out the outside world. By contrast, numerate people engage with the world and its diverse contexts and situations.

The Quantitative Literacy Design Team (2001, pp. 17–18) eloquently expressed this difference as follows: 'Mathematics climbs the ladder of abstraction to see, from sufficient height, common patterns in seemingly different things. Abstraction is what gives mathematics its power; it is what enables methods derived from one context to be applied in others. But abstraction is not the focus of numeracy. Instead, numeracy clings to specifics, marshalling all relevant aspects of setting and context to reach conclusions'.

> **Review and reflect 1.7**
>
> 1. Review your responses to the 'What is numeracy?' task in Review and reflect 1.1.
> 2. Based on your reading of this chapter, write your own definition of numeracy and how it differs from (or is similar to) mathematics.
> 3. Ask some teachers and some non-teachers what they think it means to be numerate and compare their perceptions of numeracy with your own.

WHY NUMERACY MATTERS

Being numerate involves more than mastering basic mathematics, because numeracy connects the mathematics learned at school with out-of-school situations that additionally require problem-solving, critical judgement and making sense of the non-mathematical context. It has long been known that poor numeracy has a devastating impact on young people's life chances and on the economic and social fabric of the nation. This is an issue that has been documented by the Longitudinal Surveys of Australian Youth, an initiative that follows cohorts of young Australians over a ten-year period as they transition through school to further study and work. An early report found that young people with low numeracy and literacy skills at age 4 (measured from reading comprehension and numeracy tests at school) were less likely to stay at school to Year 12 and gain entry to university or vocational education courses and more likely to experience long-term unemployment or to work in low-paid manual and labouring occupations. The report concluded that, although raising levels of literacy and numeracy would not guarantee better employment outcomes for young people, schools still had an important role to play in improving their access to post-school pathways and a secure future (Lamb 1997). More recent data confirm the pattern of disadvantage arising from poor numeracy. Among Australian respondents to the OECD's 2011–12 Programme for the International Assessment of Adult Competencies (PIAAC), there was a positive relationship between numeracy proficiency and labour force participation and employment: 36 per cent of employed people achieved a Level 3 numeracy score, compared with 23 per cent of people out of the labour force. The respective results for Levels 4 and 5 (i.e., the highest levels) were 16 per cent and 7 per cent. Those out of the labour force had the highest proportion assessed at Level 1 (the lowest level; Australian Bureau of Statistics 2014).

Parsons and Bynner (2005) were able to separate the effects of literacy and numeracy on young people's ability to function in adult life. They reported on two major longitudinal studies in the United Kingdom that followed the same groups of people from birth into adulthood, with one group born in 1958 and the other in 1970, comprising about 17,000 babies born in a particular week. Using sophisticated statistical techniques, they estimated the relative impacts of numeracy and literacy on a range of life outcomes at age 30, including 'home and family life, health, psychological well-being and political and social participation' (p. 10). Parsons and Bynner found that poor numeracy, more than low levels of literacy, severely limited successful transitions from school and subsequent work opportunities, contributing to low self-esteem, poor health prospects and lack of interest in politics or voting. While for men there was little difference between the effects of poor numeracy alone and poor literacy combined with poor numeracy, women experienced much greater negative impact from poor numeracy regardless of their level of literacy.

NUMERACY AS A CROSS-CURRICULAR RESPONSIBILITY

Achieving an adequate level of numeracy is a basic right of all students as they finish their compulsory schooling. But who should be responsible for helping young people develop their numeracy capabilities? The Quantitative Literacy Design Team (2001) maintained that for numeracy to be useful to students it must be learned in multiple contexts and in all school subjects, not just mathematics. This is a challenging notion, but the review of numeracy education undertaken by the Australian government (Council of Australian Governments 2008, p. 7) concurred, recommending 'that all systems and schools recognise that, while mathematics can be taught in the context of mathematics lessons, the development of numeracy requires experience in the use of mathematics beyond the mathematics classroom, and hence requires an across the curriculum commitment'.

How have Australian teachers and teacher educators responded to this challenge? Chapter 2 explores the process by which numeracy became formally recognised as an across-the-curriculum commitment in Australian schools and the incorporation of knowledge of numeracy teaching strategies into the Australian Professional Standards for Teachers (AITSL 2017). But for now we can point to research that has investigated Australian teachers' orientations to numeracy as they attempted to engage with numeracy across the curriculum. In a project conducted in the Australian Capital Territory, Thornton and Hogan (2004) worked with primary school and secondary school non-mathematics teachers to identify numeracy opportunities in the subjects they taught and to stimulate professional discussion on numeracy within all curriculum areas. Each teacher developed a classroom-based action research project to investigate their students' numeracy, and the group of teachers met regularly to share their observations.

Thornton and Hogan (2004) were able to identify three orientations towards numeracy, as idealised examples rather than descriptions of specific teachers. The 'separatist' recognised that mathematical skills were important but believed that teaching these skills was the responsibility of the mathematics teacher: numeracy was 'not their job'. The 'theme-maker' recognised links between mathematics and other subjects and the real world and developed units of work that integrated mathematics with other learning areas based on a theme thought to be relevant and interesting to students. For example, the theme of 'exploring the universe' drew together teaching and learning in science, social science, art, English and mathematics. This approach has been common in the primary and middle years of schooling, but it risks diminishing the important epistemological distinctions between the different learning areas in the school curriculum. The 'embedder' recognised the quantitative elements embedded in all learning areas, understood how to use mathematics within their own subject and believed that every teacher was a teacher of numeracy without necessarily being an expert mathematician. Thornton and Hogan argued that all teachers need to become willing to engage in mathematical thinking within their learning areas and to identify and capitalise on numeracy opportunities to enrich students' learning. This will require that teachers develop confidence and a disposition to embed numeracy in all areas of the curriculum.

> ### Review and reflect 1.8
>
> With which one of Thornton and Hogan's (2004) three numeracy orientations do you most identify? Explain your position to a colleague.

Professional development projects such as those conducted by Thornton and Hogan (2004) and Goos, Geiger and Dole (2011) can help teachers develop numeracy knowledge and dispositions. Pre-service teacher education should also play an important role in preparing new teachers to know, understand and implement numeracy strategies in their teaching areas. However, beginning teachers often lack confidence in their ability to foster students' numeracy within their subject areas. Milton et al. (2007) conducted an Australia-wide survey of beginning secondary school teachers to investigate how well prepared they felt to incorporate literacy and numeracy into their teaching. The respondents were 303 teachers in their first or second year of teaching, from all states and territories.

Almost all (90 per cent) saw themselves as teachers of literacy, and a clear majority (62 per cent) considered they were well prepared to teach it. However, only a little over half (55 per cent) of these new teachers saw themselves as teachers of numeracy, and around one-third perceived they were well prepared to teach it. While these negative reactions to the preparation provided by pre-service teacher education programs are worrying, some caution is needed in interpreting the survey findings. Some survey items aimed to investigate how well the respondents' courses had prepared them to use specific numeracy strategies or activities, but these items seemed instead to refer to generic teaching strategies, such as use of group work and higher order questioning. Other parts of the survey asked how well the pre-service courses had developed respondents' own numeracy concepts and skills and preparation to teach them, focusing on planning, assessment, algebra, chance and data, space, measurement and number. Yet these concepts are more recognisable as elements of the mathematics curriculum than as aspects of numeracy that support learning in other curriculum areas.

In our numeracy education research we developed an alternative way to evaluate the numeracy preparedness of teachers, in the form of the self-assessment surveys shown in Tables 1.2 and 1.3 (Goos, Geiger & Dole 2014). The surveys address three domains:

- 'professional knowledge': incorporating knowledge of students and their numeracy learning needs, knowledge of numeracy appropriate to the year levels and subjects they teach and knowledge of how to support students' numeracy learning
- 'professional attributes': incorporating personal attributes such as high expectations for students' numeracy development, a

commitment to personal professional development in order to enhance personal numeracy knowledge and teaching strategies, and acceptance of community responsibilities in communicating informed views about numeracy
- 'professional practice': incorporating establishment of supportive and challenging numeracy learning environments, planning for numeracy learning in all curriculum areas, demonstrating effective numeracy teaching strategies and using assessment strategies that allow all students to demonstrate their numeracy knowledge.

Separate surveys were developed for teachers of mathematics (primary school teachers and secondary school mathematics teachers) and teachers of disciplines other than mathematics (specialist teachers of subjects other than mathematics in primary and secondary schools).

Review and reflect 1.9

1 Select the survey relevant to your own professional circumstances—either a teacher of mathematics or a teacher of disciplines other than mathematics.
2 Annotate the survey to indicate statements that you feel you understand and indicate your level of confidence in your ability to demonstrate the capabilities described by each statement.
3 Revisit and update your annotations as you work through this book, to record your developing understanding and confidence in embedding numeracy into the subjects you teach.

Table 1.2 Numeracy self-assessment survey for all teachers of mathematics

Sub-domain	In my teaching practice, I am able to...	Rating 5-1
Professional knowledge		
Students	Understand the diversity of mathematical abilities and numeracy needs of learners	
Numeracy	Exhibit sound knowledge of mathematics appropriate for teaching my students	
	Understand the pervasive nature of numeracy and its role in everyday situations	
	Demonstrate relevant knowledge of the central concepts, modes of inquiry and structure of mathematics	
	Establish connections between mathematics topics and between mathematics and other disciplines	
	Recognise numeracy learning opportunities across the curriculum	
Students' numeracy learning	Understand contemporary theories of how students learn mathematics	
	Possess a repertoire of contemporary, theoretically grounded, student-centred teaching strategies	
	Demonstrate knowledge of a range of appropriate resources to support students' numeracy learning	
	Integrate information and communication technologies to enhance students' numeracy learning	

Sub-domain	In my teaching practice, I am able to . . .	Rating 5–1
Professional attributes		
Personal attributes	Display a positive disposition to mathematics and to teaching mathematics	
	Recognise that all students can learn mathematics and be numerate	
	Exhibit high expectations for my students' mathematics learning and numeracy development	
	Exhibit a satisfactory level of personal numeracy competence for teaching	
Personal professional development	Demonstrate a commitment to continual enhancement of my personal numeracy knowledge	
	Exhibit a commitment to ongoing improvement of my teaching of mathematics	
	Demonstrate a commitment to collaborating with teachers of disciplines other than mathematics to enhance numeracy teaching and learning	
Community responsibility	Develop and communicate informed perspectives of numeracy within and beyond the school	
Professional practice		
Learning environment	Promote active engagement in numeracy learning	
	Establish a supportive and challenging numeracy learning environment	
	Foster risk-taking and critical inquiry in numeracy learning	

Sub-domain	In my teaching practice, I am able to . . .	Rating 5–1
Planning	Highlight connections between mathematics topics and between mathematics and other disciplines	
	Cater for the diversity of mathematical abilities and numeracy needs of learners	
	Determine students' learning needs in numeracy to inform planning and implementation of learning experiences	
	Embed thinking and working mathematically in numeracy learning experiences	
	Plan for a variety of authentic numeracy assessment opportunities	
Teaching	Demonstrate a range of effective teaching strategies for numeracy learning	
	Utilise multiple representations of mathematical ideas in mathematics and in other curriculum areas	
	Sequence mathematical learning experiences appropriately	
	Demonstrate an ability to negotiate mathematical meaning and model mathematical thinking and reasoning	
Assessment	Provide all students with opportunities to demonstrate their numeracy knowledge	
	Collect and use multiple sources of valid evidence to make judgements about students' numeracy learning	

Note: 'All teachers of mathematics' includes non-specialist teachers in the early years and primary years, as well as mathematics specialist teachers in the middle years and senior years. Ratings: 5 very confident, 4 confident, 3 unsure, 2 unconfident, 1 very unconfident.

Source: Adapted from Goos, Geiger & Dole (2014).

Table 1.3 Numeracy self-assessment survey for teachers of disciplines other than mathematics

Sub-domain	In my teaching practice, I am able to ...	Rating 5-1
Professional knowledge		
Students	Recognise the numeracy knowledge and experiences that learners bring to my classroom	
	Understand the diversity of numeracy needs of learners	
Numeracy	Understand the pervasive nature of numeracy and its role in everyday situations	
	Understand the meaning of numeracy within my curriculum area	
	Recognise numeracy learning opportunities and demands within my curriculum area	
Students' numeracy learning	Demonstrate knowledge of a range of appropriate resources and strategies to support students' numeracy learning in my curriculum area	
Professional attributes		
Personal attributes	Display a positive disposition to supporting students' numeracy learning within my curriculum area	
	Recognise that all students can be numerate	
	Exhibit high expectations of my students' numeracy development	
	Exhibit a satisfactory level of personal numeracy competence for teaching	

Sub-domain	In my teaching practice, I am able to...	Rating 5–1
Personal professional development	Demonstrate a commitment to continual enhancement of personal numeracy knowledge	
	Exhibit a commitment to ongoing improvement of my teaching strategies to support students' numeracy learning	
	Demonstrate a commitment to collaborating with specialist teachers of mathematics to enhance my own numeracy learning and numeracy teaching strategies	
Community responsibility	Develop and communicate informed perspectives of numeracy within and beyond the school	
Professional practice		
Learning environment	Promote active engagement in numeracy learning within my own curriculum context	
	Establish a supportive and challenging learning environment that values numeracy learning	
Planning	Take advantage of numeracy learning opportunities when planning within my own curriculum context	
	Display willingness to work with specialist teachers of mathematics in planning numeracy learning experiences	
	Determine students' learning needs in numeracy to inform planning and implementation of learning experiences	
Teaching	Demonstrate effective teaching strategies for integrating numeracy learning within my own curriculum context	
	Model ways of dealing with numeracy demands of my curriculum area	

Sub-domain	In my teaching practice, I am able to...	Rating 5–1
Assessment	Provide all students with opportunities to demonstrate numeracy knowledge within my curriculum area	

Note: 'Teachers of disciplines other than mathematics' includes specialist teachers in the early years and primary years, as well as teachers of curriculum areas other than mathematics in the middle years and senior years. Ratings: 5 very confident, 4 confident, 3 unsure, 2 unconfident, 1 very unconfident.

Source: Adapted from Goos, Geiger & Dole (2014).

CONCLUSION

This chapter has introduced the following numeracy messages:

- Numeracy is more than number; its foundations encompass all areas of mathematics.
- Numeracy is more than 'the basics'; it involves problem-solving, reasoning, positive dispositions and a critical orientation to mathematics and how it is used in the world.
- Digital technologies are ubiquitous in the world outside school and should be exploited intelligently to develop and support students' numeracy practices.

RECOMMENDED READING

Geiger, V., Goos, M. & Forgasz, H., 2015, 'A rich interpretation of numeracy for the 21st century: A survey of the state of the field', *ZDM Mathematics Education*, vol. 47, no. 4, pp. 531–48

Goos, M., Geiger, V. & Dole, S., 2014, 'Transforming professional practice in numeracy teaching', in Y. Li, E. Silver & S. Li (eds), *Transforming*

Mathematics Instruction: Multiple approaches and practices, pp. 81–102, New York: Springer

Jorgensen Zevenbergen, R., 2011, 'Young workers and their dispositions towards mathematics: Tensions of a mathematical habitus in the retail industry', *Educational Studies in Mathematics*, vol. 76, pp. 87–100

Quantitative Literacy Design Team, 2001, 'The case for quantitative literacy', in L. Steen (ed.), *Mathematics and Democracy: The case for quantitative literacy*, pp. 1–22, Princeton, NJ: National Council on Education and the Disciplines

Thornton, S. & Hogan, J., 2004, 'Orientations to numeracy: Teachers' confidence and disposition to use mathematics across the curriculum', in M.J. Hoines & A.B. Fuglestad (eds), *Proceedings of the 28th Conference of the International Group for the Psychology of Mathematics Education*, vol. 4, pp. 313–20, Bergen: PME

2

Towards numeracy across the curriculum

In this chapter, we explore the development of numeracy as a concept and as an educational focus in Australian schooling. We begin with an overview of the history of numeracy as a national goal of schooling in the period 1989–2008, noting that these aspirational goals were articulated before there was a commonly accepted definition of what numeracy is. We then examine emerging efforts to define numeracy, together with early Australian research into improving numeracy learning and teaching. This is followed by an account of subsequent educational policy developments that supported the idea of numeracy as an across-the-curriculum responsibility. We also consider the political agenda around how numeracy is defined and assessed, with implications for practice and initial teacher education.

NUMERACY AS A NATIONAL GOAL OF SCHOOLING

Numeracy has been identified within Australia's national goals for schooling since the Hobart Declaration on Schooling, released in 1989 (Australian Education Council 1989). This report included ten agreed goals for schooling in Australia, endorsed by the state, territory and Australian ministers for education. Reference to numeracy was located in Goal 6, which specified a number of sub-goals relating to learning areas. The first two sub-goals stated the aims 'to develop in students':

a. the skills of English literacy, including skills in listening, speaking, reading and writing;
b. skills of numeracy, and other mathematical skills.

At the July 1996 MCEETYA meeting, the following goal was added: That every child leaving primary school should be able to read, write, spell and communicate at an appropriate level. In March 1997, this goal was amended to include numeracy (Australian Education Council 1997, p. ix).

In 1999, ten years after the Hobart declaration, the Adelaide Declaration of National Goals for Schooling in the Twenty-First Century was released (MCEETYA 1999). The previous ten goals were condensed to three and focused on developing 'the talents and capacities' of students, curriculum provision and 'socially just' schooling. A total of 17 sub-goals served to elaborate these three national goals. Reference to numeracy was located in Goal 2, with Sub-goal 2.2 stating that 'students should have attained the skills of numeracy and English literacy, such that every student should be numerate, able to read, write, spell and communicate at an appropriate level'.

Nine years after the Adelaide declaration, the Melbourne Declaration on Educational Goals for Young Australians proposed just two goals for schooling, which focused on schooling of excellence and for equity and promoting 'successful learners', 'confident and creative individuals' and 'active and informed citizens' (MCEETYA 2008). It was in the second goal that reference to numeracy was made and linked to developing successful learners who 'have the essential skills in literacy and numeracy and are creative and productive users of technology, especially ICT [information and communication technology], as a foundation for success in all learning areas' (p. 8). In the declaration's commitment to action, it was stated that 'the curriculum will include a strong focus on literacy and numeracy skills' (p. 13).

Although the goals for schooling have been expressed in different ways over a 20-year period, it is interesting to reflect on the positioning of numeracy within these goals, and especially on the consistent reference to skills. The use of the word *skills* tends to reduce the concept of numeracy to something that is finite, tangible and easily learned with practice. If this were the case, then numeracy skills could be measured and students' performance reported in terms of those skills. Indeed, this reductionist view of numeracy would seem to align with Australia's National Assessment Program—Literacy and Numeracy (NAPLAN) introduced in 2008, in which students' numeracy outcomes at specific junctures in schooling (i.e., in Years 3, 5, 7 and 9) are tested and reported. We return to a closer examination of numeracy assessment in Chapter 8. For now, we note that a skills-focused perspective on numeracy has been actively resisted by educators and researchers over many decades.

> **Review and reflect 2.1**
>
> 1. Access and read the common and agreed national goals for schooling in Australia published in each of the following documents: *The Hobart Declaration on Schooling (1989)* (Australian Education Council 1989), *The Adelaide Declaration of National Goals for Schooling in the Twenty-First Century* (MCEETYA 1999) and the *Melbourne Declaration on Educational Goals for Young Australians* (MCEETYA 2008). What changes in emphasis on the use of the word numeracy are evident in the statements?
> 2. Write a short description of a change in emphasis over time in the way numeracy is positioned as a goal for schooling. In particular, consider how numeracy has shifted from being linked specifically with the mathematics curriculum to being placed as a generic skill or general capability.
> 3. What is the place of both literacy and numeracy within the goals of schooling?
> 4. Share your writing with a partner who has completed this task and compare the changes and shifts you identified.

DEFINING NUMERACY

Despite the emphasis on numeracy in the Hobart, Adelaide and Melbourne Declarations, in none of these documents was numeracy ever defined. The groundwork for a definition was laid much earlier in the United Kingdom, after which further work on operationalising numeracy continued in Australia.

Numeracy in the United Kingdom

As we outlined in Chapter 1, the first use of the term *numeracy* can be traced to the United Kingdom and the Crowther Report (Ministry of Education 1959). There, it appeared under the heading 'Literacy and "numeracy"' (p. 268), with the use of inverted commas attesting to the newness of the term. In that section of the report, the authors presented a lengthy discussion of the school curriculum, emphasising that students who study science and mathematics in the final year of school should also become more 'literate' and, similarly, that those studying arts should become more 'numerate'. In introducing the word *numerate* the authors asked 'if we may coin a word to represent the mirror image of literacy' (p. 269). Thus, the Crowther Report was responsible for introducing two new words to the English vocabulary: numeracy and numerate. The report elaborated on the importance of continuing the development of students' literacy in the sciences, and numeracy in the arts, in their final year of schooling—an acknowledgement of the need for discipline studies to focus on more than content knowledge. This can also be interpreted as an early call for literacy and numeracy to be regarded as general capabilities in the school curriculum.

Another influential UK report on the teaching and learning of school mathematics, the Cockcroft Report, published in 1982, acknowledged that 'the concept of numeracy and the word itself were introduced in the Crowther Report published in 1959'. In summarising submissions to their inquiry, the authors of the Cockcroft Report cautioned against adopting narrow definitions of numeracy that had been presented to them: 'If we are to equate numeracy with an ability to cope confidently with the mathematical demands of adult life, this definition is too restricted because it refers only to ability to perform basic arithmetic operations and not to ability to make use of them with confidence in practical everyday situations' (Cockcroft 1982, pp. 10, 11). In describing what was believed to be important to the definition of numeracy and to being numerate, the Cockcroft Report provided the following:

> *We would wish the word 'numerate' to imply the possession of two attributes.* The first of these is an 'at-homeness' with numbers and an ability to make use of mathematical skills which enables an individual to cope with the practical mathematical demands of his everyday life. The second is an ability to have some appreciation and understanding of information which is presented in mathematical terms, for instance in graphs, charts or tables or by reference to percentage increase or decrease. Taken together, these imply that a numerate person should be expected to be able to appreciate and understand some of the ways in which mathematics can be used as a means of communication. (p. 11, original emphasis)

Numeracy in Australia

Australian researchers and organisations were beginning to argue for similarly expansive definitions of numeracy and what it means to be numerate. Speaking at a *literacy* conference, Sue Willis (1990a, p. 84), an Australian mathematics education researcher, proposed that 'being numerate, at the very least, is about being able to use mathematics—at work, at home and for participation in community or civic life'. She went on to explain that numeracy was 'not about the acquisition of even a large number of decontextualised mathematical facts and procedures ... neither is it about learning mathematics for its own sake ... and it is certainly not about getting through selection processes ... Numeracy is about practical knowledge where practical should not be confused with low level "hands on" or procedural knowledge'.

In another forum, Willis (1990b, p. vii) claimed that being numerate meant being able 'to function effectively mathematically in one's everyday life, at home and at work'. In later documents building on this work (e.g., Morony et al. 2004; Thornton & Hogan 2004), Willis's definition of numeracy was expanded to include reference to the affective domain in the form of dispositions: 'Being numerate, at the very least, is about having the competence and disposition to use mathematics to meet the general demands of life at home, in paid work, and for participation in community and civic life' (Willis 1992, cited in Thornton & Hogan 2004, p. 313).

Robyn Zevenbergen is another Australian mathematics education researcher who has shaped our current conceptions of numeracy. In 1995, she contrasted three perspectives on numeracy—functional, cultural and socially critical—in arguing for the importance of developing a more broadly encompassing definition of the concept. Zevenbergen described 'functional numeracy' as being associated with the development of basic skills, particularly number facts and operations as skills necessary for survival in the wider society. According to Zevenbergen, this view of numeracy derived from a rigid perspective on what mathematics is and what it is used for and from a narrow vision of what mathematics is and what it is not. Such a view appears to focus predominantly on algorithms and correct procedures, ignoring personal intuitive mathematics.

Zevenbergen suggested that 'cultural numeracy' recognised and valorised the cultural specificity of mathematics and ensured that it valued the personal knowledge and mathematics practised by a particular group. As a consequence, numeracy may look different in different cultures. Finally, she described 'socially critical numeracy' as using mathematics as a tool for critique. This perspective emphasises the capacity to critically examine the ways in which mathematics is used to represent information and exert a power relationship. According to Zevenbergen, socially critical numeracy is numeracy as empowerment. Her analysis highlighted and cautioned against adopting definitions or perspectives of numeracy associated only with attainment of basic skills.

While researchers and educators in Australia and the United Kingdom were strongly arguing against narrow, skills-based definitions of numeracy, similar debates were occurring over definitions of *literacy* and what it means to be *literate*. In both areas, there was opposition to narrow perspectives of literacy and numeracy as sets of basic skills. Cumming (1996, p. 11), a noted literacy educator, agreed that 'numeracy, paralleling early notions of literacy, was for a long time interpreted as a back-to-basics arithmetic and computation work'. It was at this time that the term *multiliteracies* was coined (New London Group 1996) to reflect new times of globalisation and digital technologies and to continue to argue for a more broadly encompassing perspective of what literacy and numeracy mean. As multiliteracies took account of literacy in an ever-evolving and rapidly changing technological and information-driven age, it was recognised that the definition of numeracy would continuously evolve and expand (Noss 1998) and that numeracy should be regarded as an 'elastic term' (Doig 2001). Cumming (1996, p. 13) noted that 'the history of constructions of numeracy can thus be traced in parallel with the changing constructions of literacy'.

The period from the late 1990s to the early 2000s in Australia was a time of considerable debate and discussion about schooling, and much energy was expended on the development of a definition of numeracy. These efforts occurred at the time when the Ministerial Council on Education, Employment, Training and Youth Affairs (MCEETYA) extended the ten national common and agreed goals of schooling to include a specific focus on literacy and then on numeracy. A major forum for discussing what it meant to be numerate was provided by the Numeracy Education Strategy Development Conference (DEETYA 1997), which was held in Western Australia in May 1997. This conference, funded by the Commonwealth through the Department of Employment, Education, Training and Youth Affairs, was orchestrated by the Education Department of Western Australia and the Australian Association of Mathematics Teachers. It brought together community representatives to deliberate 'on the current state of numeracy education in Australia and how best to move forward'. One of the driving forces behind this conference was the conviction that 'definitions or descriptions of what constitutes numeracy need to be developed, elaborated, exemplified, discussed and revised' (DEETYA 1997, p. 1).

At this conference, various state and territory jurisdictions put forward descriptions of numeracy that were currently in their curriculum documents. For example, the following definition of numeracy was presented as part of the Education Department of Tasmania's 1994 numeracy strategy: 'To be numerate is to have and be able to use appropriate mathematical knowledge, understanding, skills, intuition and experience whenever they are needed in everyday life. Numeracy is more than just being able to manipulate numbers. The content of numeracy is derived from the five strands of the mathematics curriculum—space, number, measurement, chance and data, (pattern and) algebra' (cited in DEETYA 1997, p. 3). Similarly, the Department of Education Queensland's Literacy and Numeracy Strategy 1994–98 included the following definition: 'Numeracy involves abilities which include interpreting, applying and communicating mathematical information in commonly encountered situations to enable full, critical and effective practice in a wide range of life roles' (cited in DEETYA 1997, p. 2). While other definitions from published literature were presented and key themes, notions and concepts extracted and debated, the conference report made a recommendation to MCEETYA to adopt the following position on numeracy: 'To be numerate is to use mathematics effectively to meet the general demands of life at home, in paid work, and for participation in community and civic life' (p. 15).

In 2000 the Australian Department of Education, Training and Youth Affairs published a national numeracy strategy that set goals for literacy and numeracy and adopted the definition of numeracy recommended by the Numeracy Education Strategy Development Conference of 1997. This definition, while succinct and widely accepted, was accompanied by a longer statement that described what it means to be numerate, noted the centrality of numeracy in the school curriculum and offered a summary of the concepts, skills, knowledge and dispositions that characterise a numerate person. For completeness, we reproduce this statement about numeracy and being numerate that was presented to Australia as a result of the 1997 conference:

To be numerate is to use mathematics effectively to meet the general demands of life at home, in paid work, and for participation in community and civic life.

In school education, numeracy is a fundamental component of learning, performance, discourse and critique across all areas of the curriculum. It involves the disposition to use, in context, a combination of:

- underpinning mathematical concepts and skills from across the discipline (numerical, spatial, graphical, statistical, and algebraic);
- mathematical thinking and strategies;
- general thinking skills; and
- grounded appreciation of context. (DEETYA 1997, p. 15)

NUMERACY RESEARCH AND DEVELOPMENT

In the years 2001 to 2004, the Australian government invested $7 million in the Numeracy Research and Development Initiative. The purpose of the initiative was to 'investigate a range of teaching and learning strategies that lead to improved numeracy outcomes', with projects ranging from those that investigated 'broad issues such as school policy and pre-service teacher education to more specific factors influencing numeracy outcomes such as the physical and social structure of classrooms' (DEST 2005, pp. 3, 4). Various projects were implemented across states and territories and involved university researchers working with education sectors.

One major project funded by this initiative was the Numeracy Across the Curriculum Project, led by John Hogan (DEST 2004a,

2004b). This project operationalised the definition of numeracy as being more than skills and procedures and provided groundbreaking research evidence that mathematics is necessary but not sufficient for numeracy (Hogan 2000a). The aim of the project was to support teachers in developing a way of thinking about numeracy as the capacity to apply mathematical, contextual and strategic know-how in context.

Hogan's Numeracy Across the Curriculum Project built on the work of Willis (1998), who identified three aspects of numeracy and the types of know-how associated with each:

- 'Numeracy as mathematics': this is the basic skills view, which also dominates in schools. Mathematical know-how is developed by increasing students' mathematical repertoire or knowledge.
- 'Numeracy as communicative competence': this view acknowledges that mathematics is embedded in everyday situations and numeracy is context specific. Contextual know-how is developed by increasing students' repertoire of situations or practical contexts in which they need to use mathematics for a specific purpose. While this perspective is common in adult education, it is difficult for schools to provide authentic contexts that match those in which numeracy skills are exercised in everyday life.
- 'Numeracy as strategic mathematics': this view emphasises the importance of mathematical processes, applications and dispositions in choosing and using mathematical skills in the service of non-mathematical goals. Strategic know-how is developed by increasing students' repertoire of strategies for dealing with unfamiliar problems.

Being numerate involves *all three* of these interpretations. Strategic know-how may be particularly important as a potential bridge between school mathematics (content focus) and mathematics as practised in the real world (context focus; see Figure 2.1).

This way of thinking about numeracy also incorporated three roles: the 'fluent operator' (a fluent user of mathematics in familiar settings), the 'learner' (a capacity for deliberate use of mathematics to learn) and the 'critic' (a capacity to be critical of the mathematics chosen and used) (DEST 2004a). The three roles provided the numeracy framework for Hogan's project.

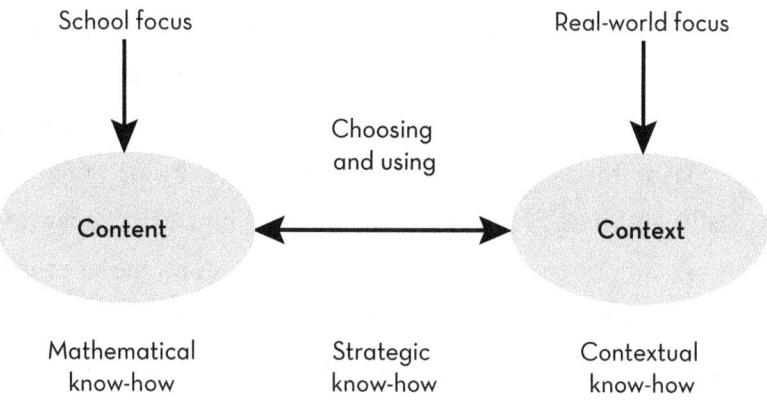

Figure 2.1 Three kinds of know-how for being numerate
Source: Board of Teacher Registration, Queensland 2005, p. 1.

Morony et al. (2004, p. 6) provided an illustration of how this numeracy framework was applied in the Numeracy Across the Curriculum Project. They described a science classroom where children were watching a video about blood that made reference to 250 million blood cells. This number was something that the children found difficult to conceptualise. One boy asked:

> What does 250 million look like? That is a lot of blood cells! How is it possible for all of those cells to fit into our body?
>
> If you were fatter, like myself, wouldn't you have more red blood cells?

The teacher had incorrectly assumed that her students would be fluent operators who knew how to represent and visualise this quantity. However, the student who asked the questions assumed the role of learner, trying to make sense of how so many things could fit inside one's body (however large that might be). As the class investigated other information sources, they found conflicting information about the number of blood cells in the body. Morony et al. (2004, p. 6) explained that 'having the capacity to ask questions about whether the mathematics is appropriate, and to try to reconcile information from different sources, is a key aspect of being critically numerate'. This project tested the numeracy framework in a range of primary school classrooms. As a result, the researchers noted that teachers increasingly became more aware of the numeracy demands of subjects other than mathematics and were able to diagnose students' numeracy needs. Teachers also found the numeracy framework assisted them in planning and designing learning experiences so that students had opportunities to 'intelligently apply mathematics in a situation'.

NUMERACY IN EDUCATIONAL POLICY

Collectively, the projects undertaken through the Numeracy Research and Development Initiative were analysed and the outcomes collated into a summary of evidence-based approaches

Review and reflect 2.2

The Numeracy Across the Curriculum Project described above developed a set of questions used by a person acting numerately to help teachers understand their students' numeracy learning. The questions bring together the three roles of 'fluent operator', 'learner' and 'critic' (DEST 2004a) across the three domains of mathematical, contextual and strategic knowledge (Table 2.1).

Table 2.1 Questions asked by a person acting numerately across knowledge domains

Mathematical domain	Contextual domain	Strategic domain
Learner		
• What mathematical ideas and techniques are being used, or could be used, in this situation? • Do I understand them? • Do I need to understand them better? • How will I use them?	• What is going on here? • How is this context impacting on the mathematics? • How will the mathematics impact on this context? • What else about this context do I need to find out about? • What adaptations do I need to make to the mathematics in order to apply it to this setting and fit the constraints of the situation?	• Have I asked if mathematics might help? • Have I found out about what mathematics I could use? • What assumptions have to be made in order to use this mathematics? • Have I asked for support from others and found the necessary information where necessary?

Mathematical domain	Contextual domain	Strategic domain
	• How accurate do I need to be here? • Does my answer make sense in this situation?	• Have I taken account of the constraints in the situation and adapted the mathematics to suit? • Have I made a decision about how accurate I need to be? • Have I checked to see if my answer makes sense in this situation?
Critic		
• Does the mathematics make sense? • Can I justify using this mathematics for this purpose? • What other mathematics could I use?	• Does it make sense to use the mathematics in this situation? • How is the mathematics in this situation being used? • Who, in this setting, is using this mathematics? • Why are they using it in this way? • Who are they using it for?	• Have I checked to see if the mathematics being used by others is appropriate? • Have I understood the assumptions being made by people using this mathematics?

Mathematical domain	Contextual domain	Strategic domain
	• Are the outcomes and conclusions being made in this situation justified and relevant?	• Have I evaluated the mathematics used to see if it was the best way?

Fluent operator

A numerate person is smooth and almost automatic in their use of mathematics in familiar, everyday situations. This is the comfortable, quick and ready, almost unconscious use of mathematics, including with self-taught, informal methods— the mathematics of 'just plain folks'.

Source: DEST 2004b, appendix 10.

1. Use the questions in Table 2.1 to analyse your experience of responding to the cake tin and weedkiller numeracy tasks in Chapter 1.
   - Which questions are closest to how you were thinking about these tasks?
   - To what extent did you take on the roles of 'fluent operator', 'learner' and 'critic'?
   - In which knowledge domains were you operating?
2. Compare and discuss your analysis with that of a colleague. What similarities and differences do you see in your responses to these numeracy tasks?

that enhance students' numeracy capabilities (DEST 2005). One of the main messages from the research undertaken through this initiative was the imperative to clarify the difference between numeracy and mathematics in the minds and actions of all stakeholders, as it was repeatedly found that the term numeracy was 'viewed by many people, including teachers, as synonymous with school mathematics skills' (p. 29). However, despite the significant investment by the Australian government in these projects, and the examples of how to embed numeracy across the curriculum provided by the research of Hogan and colleagues, this idea was not taken up on a large scale in schools. Further impetus for numeracy as a cross-curricular priority emerged as the result of the National Numeracy Review (Council of Australian Governments 2008), the development of a nationally consistent Australian Curriculum and the introduction of the Australian Professional Standards for Teachers, which underpin national accreditation of initial teacher education programs.

The National Numeracy Review

The national Numeracy Education Strategy Development Conference of 1997 resulted in the formation of a working party that aimed to establish some fundamental principles of what numeracy is and is not. These included the following:

- 'Literacy and numeracy are distinct'.
- 'Numeracy is more than number sense'.
- 'Numeracy is not a synonym for school mathematics' (DEETYA 1997, p. 11).
- 'Numeracy is cross-curricular and is a responsibility for all educators' (p. 18).

The working party emphasised that numeracy, like literacy, is 'everyone's business' (DEETYA 1997, p. 12). In the United States, the Quantitative Literacy Design Team (2001, p. 18) led by Steen also identified the importance of a cross-curricular approach to numeracy: 'To enable students to become numerate, teachers must encourage them to see and use mathematics in everything they do ... Fortunately, because numeracy is ubiquitous, opportunities abound to teach it throughout the curriculum'.

Further support for this idea came from the Council of Australian Governments' (2008, p. vii) National Numeracy Review, which aimed to provide 'an opportunity for a stocktake of research-based evidence about good practice in numeracy and the learning of mathematics'. The review was carried out in a context of transition from state-based to national assessments of literacy and numeracy. NAPLAN began in 2008 and annually tests all students in Years 3, 5, 7 and 9. The NAPLAN numeracy tests 'assess the proficiency strands of understanding, fluency, problem-solving and reasoning across the three content strands of mathematics: number and algebra; measurement and geometry; and statistics and probability' (ACARA 2017). From this definition, it is clear that these tests draw from the Australian Curriculum: Mathematics and arguably would be regarded as assessing mathematics rather than the rich conception of numeracy presented throughout this book. (We discuss numeracy assessment further in Chapter 8.)

While the National Numeracy Review acknowledged that questions needed to be resolved around the distinction between numeracy and mathematics, it is significant that its first recommendation was that 'all systems and schools recognise that, while mathematics can be taught in the context of mathematics lessons, the development of numeracy requires experience in the use of mathematics beyond the mathematics classroom, and hence requires an *across the curriculum commitment*' (Council of Australian Governments 2008, p. 7, emphasis added).

The Australian Curriculum

At around the same time as the National Numeracy Review, work was beginning on developing an Australian Curriculum in all learning areas, for use in every jurisdiction. Mathematics was one of the first four learning areas to be developed in the Foundation to Year 10 Australian Curriculum. Curriculum development work began in July 2008, leading to the release of the *National Mathematics Curriculum: Framing paper* (National Curriculum Board 2008) and then the *Shape of the Australian Curriculum: Mathematics* paper (National Curriculum Board 2009) to guide the writing of the curriculum itself. The framing paper implied that numeracy and mathematics are essentially the same. It offered a proposal to specify 'authentic proficiency standards for numeracy … as part of the mathematics curriculum' in order to avoid 'an artificial distinction between it and mathematics' (National Curriculum Board 2008, p. 4). The later *Shape* paper retreated from this position, adopting a perspective more consistent with the notion of numeracy as an overarching general capability that 'emphasizes the key role of applications and utility in learning the discipline of mathematics, and illustrates the way that mathematics contributes to the study of other disciplines' (National Curriculum Board 2009, p. 5).

The Australian Curriculum espouses the notion of numeracy as a general capability to be developed in all learning areas and offers advice within each learning area for developing curriculum-specific numeracy. (This idea is further explored in Chapter 4.) Thus, there is a clear expectation that all teachers will be responsible for developing their students' numeracy capabilities.

Australian Professional Standards for Teachers

The idea of professional standards setting out what teachers should know and be able to do had gained currency in Australia by the early 2000s, with some educational jurisdictions developing standards frameworks for teachers working in their state systems (e.g., Education Queensland 2005). In addition, the national professional associations for teachers of mathematics, science and English collaborated with university-based education researchers to develop subject-specific standards for excellence in teaching these subjects (e.g., AAMT 2006). The movement towards a national professional standards framework was initiated by the Ministerial Council for Education, Early Childhood Development and Youth Affairs (MCEEDYA) in 2009. The newly created Australian Institute for Teaching and School Leadership assumed responsibility for validating and finalising the Australian Professional Standards for Teachers in July 2010. The standards were endorsed by MCEEDYA in December 2010.

There are seven standards grouped into three domains: 'professional knowledge', 'professional practice' and 'professional engagement' (AITSL 2017; see Table 2.2). These domains represent one dimension of the standards. The second dimension identifies four career stages: 'graduate teacher', 'proficient teacher', 'highly accomplished teacher' and 'lead teacher'. Together, the two dimensions map out professional capabilities throughout a teacher's career.

Table 2.2 Australian Professional Standards for Teachers

Professional domains	Standards
Knowledge	1 Know students and how they learn
	2 Know the content and how to teach it
Practice	3 Plan for and implement effective teaching and learning
	4 Create and maintain supportive and safe learning environments
	5 Assess, provide feedback and report on student learning
Engagement	6 Engage in professional learning
	7 Engage professionally with colleagues, parents/carers and the community

Source: AITSL 2017.

For each standard, there are a number of focus areas. Within Standard 2, 'Know the content and how to teach it', there is an explicit focus on teachers' responsibility for developing their students' numeracy: 'Literacy and numeracy strategies: Know and understand literacy and numeracy teaching strategies and their application in teaching areas' (AITSL 2017, Standard 2.5).

The Australian Professional Standards for Teachers have assumed particular significance for initial teacher education as a mandatory program accreditation framework, with the Australian Institute for Teaching and School Leadership being responsible for accreditation procedures and state-based teacher registration bodies implementing these procedures. This means that universities offering initial teacher education programs must identify where each graduate teacher standard is taught, practised and assessed and must require that pre-service teachers have demonstrated successful performance against all of the graduate teacher standards prior to graduation.

> **Review and reflect 2.3**
>
> 1. In the Australian Professional Standards for Teachers (AITSL 2017), find Standard 2.5, 'Literacy and numeracy strategies'.
> 2. Click on the 'Illustrations of practice' that are linked to this standard and review their relevance to embedding numeracy across the curriculum.
> 3. Begin developing a portfolio of evidence that you know and understand numeracy teaching strategies and their application in your teaching area. Your responses to many of the Review and reflect activities in this book will provide appropriate evidence.

> **Review and reflect 2.4**
>
> 1. Create a visual representation of the development of numeracy education policy in Australia. This could be an annotated, scaled timeline, a poster, a flow chart, a concept map or something else. Your representation should show how and why definitions and understandings of numeracy have evolved over time.
> 2. Compare your representation with that of a partner and discuss the strengths and weaknesses of each.
> 3. If appropriate, revise your representation in light of feedback from your partner.

CONCLUSION

In tracing the historical development of numeracy in Australia in tandem with government policy, we can see the source of tension between two numeracy agendas. On the one hand, there is a continual struggle to promote a rich definition of numeracy as more than basic skills and to encourage teachers to engage students in rich numeracy learning experiences. On the other hand, the political agenda of measuring and reporting students' numeracy outcomes at particular school year levels may serve to perpetuate a skills-based definition of numeracy. The promotion of a rich definition of numeracy in the minds of teachers, students, school leaders and the general public is an important and ongoing educational goal.

RECOMMENDED READING

Council of Australian Governments, 2008, *National Numeracy Review Report*, <http://webarchive.nla.gov.au/gov/20080718164654/http://www.coag.gov.au/reports/index.htm#numeracy>, retrieved 11 March 2018

Department of Employment, Education, Training and Youth Affairs, 1997, *Numeracy = Everyone's Business: The report of the Numeracy Education Strategy Development Conference, May 1997*, Adelaide: Australian Association of Mathematics Teachers

Department of Education, Science and Training, 2004a, *Numeracy Across the Curriculum*, Canberra: Commonwealth of Australia

Department of Education, Science and Training, 2004b, *Numeracy Across the Curriculum: Appendices*, Canberra: Commonwealth of Australia

Department of Education, Training and Youth Affairs, 2000, *Numeracy, a Priority for All: Challenges for Australian schools*, Canberra: DETYA

Ministerial Council on Education, Employment, Training and Youth Affairs, 2008, *Melbourne Declaration on Educational Goals for Young Australians*, <www.curriculum.edu.au/verve/_resources/National_Declaration_on_the_Educational_Goals_for_Young_Australians.pdf>, retrieved 17 November 2017

Willis, S., 1990, *Being Numerate: What counts?*, Hawthorn, Vic.: Australian Council for Education Research

3

Numeracy in the 21st century

If numeracy is more than a set of basic mathematics skills needed to survive in the world, then what *are* the dimensions of numeracy—especially in the context of those capabilities needed to meet the challenges of life in the 21st century? In Chapter 1, the origins of conceptions of numeracy, how it is different from mathematics and why it matters were discussed. Chapter 2 outlined the recognition of numeracy as an educational goal and explored the development of numeracy as an across-the-curriculum commitment in Australian schools. In this chapter, we introduce a model of numeracy (Goos, Geiger & Dole 2014) designed to capture the demands of 21st century personal, civic and work life and describe its dimensions. The model was developed to provide support for teachers wishing to promote their students' numeracy capabilities in the range of learning areas found in schools (Geiger, Goos, Dole et al. 2013).

The first section of this chapter describes the five dimensions of the numeracy model. Each dimension is illustrated via examples drawn from real-world contexts. Two case studies are then presented, which illustrate how the model applies in a holistic way to problems in the real world. The chapter concludes with a discussion of the strengths and limitations of the model.

THE 21ST CENTURY NUMERACY MODEL

The 21st Century Numeracy Model brings together features of numeracy described in previous research (e.g., Hogan 2000a) while also extending the definition of numeracy to accommodate the evolving nature of knowledge, work and technology (Goos, Geiger & Dole 2010, 2014). This model was initially developed to provide a workable description of numeracy in order to support teachers' planning for effective practice across the curriculum (see Chapter 1) and to provide structure for reflection on lessons in which numeracy development was a feature.

The model consists of four core dimensions: attention to real-life *contexts*; application of *mathematical knowledge*; use of physical, representational and digital *tools*; and promotion of positive *dispositions* towards the use of mathematics to solve real-life problems. These four dimensions are embedded in a fifth dimension—a *critical orientation*—that is intertwined through and between the core dimensions. This dimension requires the appropriate selection and application of mathematics to real-world problems as well as the interpretation and critique of results. A summary of the dimensions of the numeracy model appears in Table 3.1, and the model is additionally represented in Figure 3.1. It is then described according to each of its dimensions.

Table 3.1 Dimensions of the 21st Century Numeracy Model

Dimensions	Explanations
Contexts	Use of mathematics to act in and on the world and thus in a range of real-world situations both within schools and beyond school settings
Mathematical knowledge	Concepts and skills, problem-solving strategies and estimation capacities
Dispositions	Confidence and willingness to use mathematical approaches to engage with life-related tasks; preparedness to make flexible and adaptive use of mathematical knowledge
Tools	Use of physical (e.g., models, measuring instruments), representational (e.g., symbol systems, graphs, maps, tables) and digital (e.g., computers, applications, internet) tools to mediate and shape thinking
Critical orientation	Use of mathematical information and activity to analyse and evaluate information and data within a given context, make decisions and judgements, form opinions, add support to arguments and challenge an argument or position

Source: Goos, Dole & Geiger 2012.

Contexts

Context is positioned at the centre of the model because numeracy is about the use of mathematics to act on the real world (Quantitative Literacy Design Team 2001). Consequently, the use of mathematics alone, without a context, cannot be seen as numerate activity. Thus, context is at the heart of numeracy.

The role of context becomes obvious when comparing the way mathematics is most often taught in school to its implementation when attempting to solve problems in the real world (Straesser

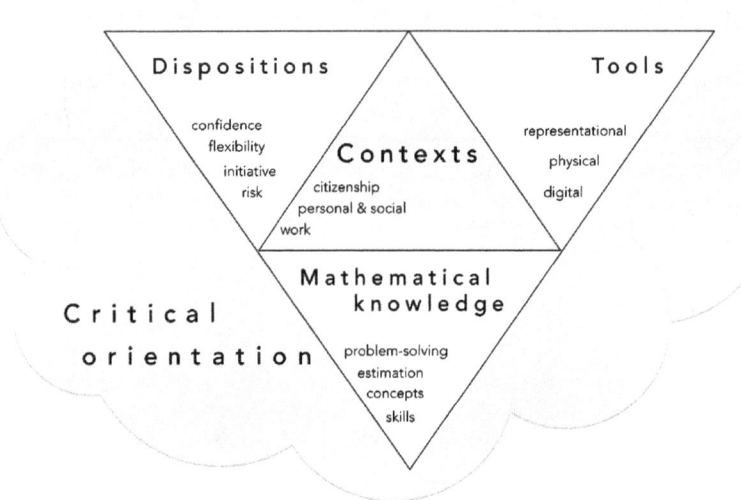

Figure 3.1 The 21st Century Numeracy Model
Source: Goos, Geiger & Dole 2010, 2014.

2007). Differences between mathematics in school and its application in real life are apparent—for example, in the workplace, where the context strongly influences the way mathematical knowledge and skills are used (e.g., Hoyles et al. 2002; Noss et al. 2000). A nurse preparing a dose of medication, for example, must often take into account such factors as the age, weight, state of health and sometimes sex of a patient. Guidelines related to these factors will need to be consulted before the appropriate and safe amount of the medication is administered. Thus, the mathematics involved is very much bound to the context and far more complex than the type of simple proportion calculation that might appear in a school textbook. The use of mathematics in this case requires judgement and adaptable thinking based on the specific characteristics or needs of the patient.

The vital role of context in applying mathematics to the real world means that a numerate person needs to be able to select and use mathematics in a manner specific to the features of a particular problem. This implies that the process of becoming more numerate is not necessarily about learning more mathematics; instead, it involves extending one's capacity to apply mathematics to an increasing range of different circumstances or contexts (Quantitative Literacy Design Team 2001).

> **Review and reflect 3.1**
>
> 1 With a partner, select a topic in mathematics you believe is used in different contexts in personal, civic or work life.
> 2 By working individually, write down the ways this topic in mathematics is used within different contexts.
> 3 Compare the list you generate with that of your partner. Together, think of uses of this mathematics in additional contexts.

Mathematical knowledge

Using mathematics to solve problems in the real world requires personal mathematical knowledge relevant to a particular situation. The mathematical knowledge needed to solve a real-world problem includes appropriate concepts and skills as well as capabilities such as making sensible estimations and applying problem-solving strategies (Jorgensen Zevenbergen 2011). This means a numerate person has both personal mathematical knowledge and the know-how to apply this knowledge in ways relevant to a specific context. Thus, becoming numerate is not just about acquiring a certain standard or level of mathematical knowledge, as it also involves making use of this knowledge in ways that make sense in a context. This means interpreting a problem from the real world in a mathematical way in order to select relevant mathematical knowledge. Further, once specific mathematical knowledge is selected and applied, it will also be necessary to check if results make sense within the context of the original problem.

Review and reflect 3.2

The drinks menu in Table 3.2 comes from a restaurant in Seoul, South Korea. Two brands of beer are listed, Max and Red Rock, along with the prices for different volumes, Regular, Jug and Pitcher. The currency used in Korea is the won, for which the symbol ₩ is used. Each Australian dollar is equal to approximately 1000 won. If you compare the prices against volumes you will see something strange.

Table 3.2 Comparison of prices of Max and Red Rock beer by volume

Size	Volume (cm³)	Price (₩)	
		Max	Red Rock
Regular	500	3500	4000
Jug	750	6000	7000
Pitcher	2000	11,000	13,000

1 Work with a partner to identify the irregularity and explain why things are not what you would expect. In doing so, write down:

> - the mathematical knowledge you applied to the problem
> - any estimation skills you used
> - the problem-solving strategies you employed.
> 2. It is uncertain if the prices on this list were part of a deliberate plan or a mistake. Present an argument for both of these possibilities.

Dispositions

To use mathematics to solve problems in the real world, a person must first want to do so and feel confident they might be successful. Because using mathematics in life-related contexts can often be challenging, it is necessary to show initiative, exercise flexible thinking and be prepared to take the risk of being unsuccessful. As many people lack confidence in their mathematical ability, such risk may prevent them from participating fully in many aspects of social, cultural and civic life where using mathematics is a vital skill (e.g., Attard et al. 2016). This has been documented as an anxiety faced by pre-service teachers as well as the general public (Wilson & Thornton 2006).

A numerate person, therefore, must be willing to engage with and then persist in the face of the inevitable challenges that result from working with problems found in the real world. This means developing a positive disposition to using mathematics when it provides a viable way of finding a solution to a real-world problem.

> **Review and reflect 3.3**
>
> 1 Read the paper by Wilson and Thornton (2006). Do you know anyone who experiences the type of mathematics anxiety described?
> 2 Discuss with a partner the effect of such anxiety on participation in personal, civic and work life.
> 3 Make a list of strategies for building teachers' positive dispositions to using mathematics within the contexts of their different subject areas to address real-world issues.

Tools

Using mathematics to solve problems in personal, civic and work life almost invariably involves the use of some type of tool (e.g., Drijvers & Weigand 2010; Hoyles et al. 2010). Tools may be physical (e.g., measuring tapes, protractors), representational (e.g., graphs, maps, tables) or digital (e.g., computers, calculators, the internet, smartphone apps).

Digital tools are almost ubiquitous in all aspects of life, and this widespread adoption of technology has dramatically changed the way mathematics is used to perform both mundane calculations and complex tasks that require initiative, exploration and novel approaches to problem-solving (Jorgensen Zevenbergen 2011). Recall, for example, the use of a spreadsheet by young workers developing a tender in Case study 1.1.

While there are clear advantages related to the use of digital tools to organise, manage and explore data and information, such activities require the ability to interpret and make judgements about the validity and appropriateness of a result produced by a digital tool (Geiger, Goos & Dole 2015; Hoyles et al. 2010). Being willing and able to form judgements about the validity of a result produced by a tool, rather than blindly accepting an output, connects this dimension to the critical orientation aspect of being numerate. A critical orientation is discussed later in the chapter.

> ### Review and reflect 3.4
>
> Who is responsible for carbon dioxide emissions? The fundamental factor in global warming has been identified by scientists as the level of human-generated carbon dioxide emissions. In debates that are concerned with this issue, questions arise over who is most responsible and who can make the most difference in curbing global warming. This is a far more complex issue than first sight might suggest. In this task, you will make use of relevant data to explore the issue of national responsibility for carbon dioxide emissions and make evidence-based suggestions about dealing with this problem.
>
> 1 Go to the Gapminder (2018b) bubble graph showing cumulative carbon dioxide emissions. Click on the start arrow below the graph and write down observations about how cumulative emissions have changed over time and which countries appear to have been the largest contributors to those emissions. These observations should make use of data in making claims. Include both Australia and Germany in the countries you comment on. Countries can be identified by checking the tick boxes on the right-hand side of the window.

2. Now go to the Gapminder (2018a) bubble graph showing carbon dioxide emissions per capita. Click on the start arrow and write down observations about how per capita emissions have changed over time and which countries appear to have been the largest contributors to those emissions. These observations should make use of data in making claims. Again, include both Australia and Germany in the countries you comment on.
3. Compare your responses for cumulative carbon dioxide emissions and carbon dioxide per capita and write a paragraph of commentary on what you have observed.
4. After completing this activity, discuss with a partner the role of the digital tool Gapminder in developing your commentary:
    - What advantages did it offer?
    - Were there any disadvantages?
    - What cautions would you advise in interpreting available results?

Critical orientation

The four previously described dimensions of numeracy are embedded in a fifth overarching dimension—critical orientation. This dimension is concerned with interpretive, evaluative and analytical aspects of being numerate.

Ernest (2002, p. 6) argues that mathematics has both a utilitarian purpose (basic skills required to function in personal life and in the workplace) and a role in social empowerment (ability to engage in dialogue about actions that affect the future).

The empowered learner will not only be able to pose and solve mathematical questions (mathematical empowerment), but also will be able to understand and begin to answer important questions relating to a broad range of social uses and abuses of mathematics (social empowerment). Many of the issues involved will not seem primarily to be about mathematics, just as keeping up to date about current affairs from reading broadsheet newspapers is not primarily about literacy. Once mathematics becomes a 'thinking tool' for viewing the world critically, it will be contributing to both the political and social empowerment of the learner, and hopefully to the promotion of social justice and a better life for all.

To view the world in a mathematically critical way is to use mathematics to form evidence-based opinions and to make judgements or decisions by considering available information and data. This requires the ability to critique in addition to the capacity to interpret mathematical results—a critical orientation.

According to the Quantitative Literacy Design Team (2001), being numerate (or what they termed 'quantitatively literate') is vital in an increasingly complex and information-drenched society. They argued that, in such a world, citizens must be quantitatively literate and capable of using mathematics to think about both commonplace issues and matters related to the common good. This includes activities which range from deciding which breakfast cereal is the best value for money at a supermarket based on price and volume to challenging authority through the use of evidence-based arguments.

The ability to be critical is crucial in a world where mathematical information is increasingly used by the media to persuade, manipulate and shape opinion about social or political issues locally, nationally and internationally (Jablonka 2015; Rosa & Clark Orey 2015). A person who has developed a critical orientation to numeracy will be less prone to such influences and will also be in a position to present alternative arguments by looking at relevant data and information in an informed way.

In a paper about developing a mathematics curriculum that meets the needs of all students, Sue Willis (2001) illustrated the importance of being numerate by drawing on an advertisement promoting a sale of a 500-gram jar of chocolate hazelnut spread that portrayed it as offering value for money. The advertising claimed, '33 per cent free! 500 grams for the price of 375 grams'.

One could easily see the claim in the advertisement as correct, as, after all, 33.3 per cent of 375 grams is 125 grams, which when added to 375 grams makes up 500 grams. However, another perspective might be that the claim is related to 33.3 per cent of 500 grams (the amount in the container), which is 166 grams. This is more than 125 grams, so the advertisement could appear to be over claiming. This example demonstrates the role of context, or how real-world situations are interpreted, in using mathematics to form opinions

Review and reflect 3.5

1 Evaluate the claim made in the chocolate hazelnut spread advertisement described in this section.
2 Compare your thinking with a partner.
3 Write a paragraph summarising any different points of view discussed and your final opinion on the issue.

or make decisions. The example also highlights the importance of being able to take a critical orientation to a commonplace issue such as getting the best value for money on purchases from a supermarket. An additional example is given in Review and reflect 3.6, continuing the chocolate theme and requiring the deployment of critical capabilities to make a judgement about what might appear to be an obvious conclusion based on the data presented.

Review and reflect 3.6

The graph shown in Figure 3.2 portrays the correlation between chocolate consumption and the number

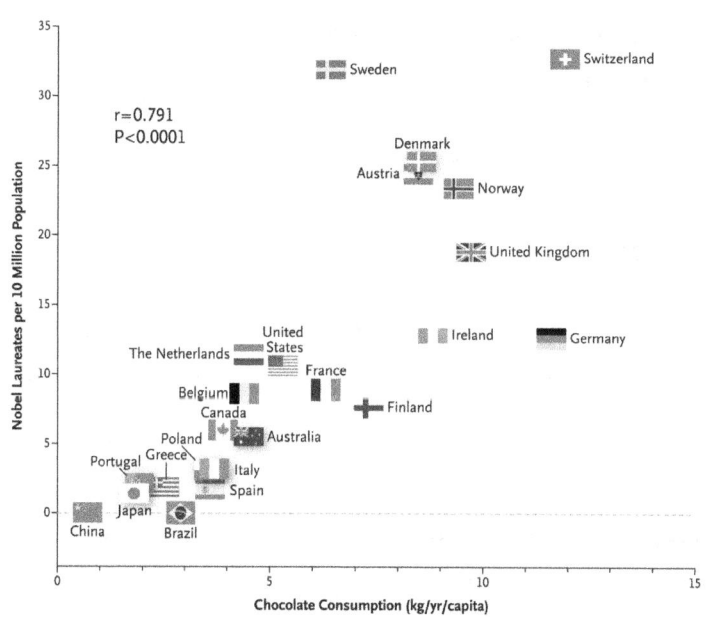

Figure 3.2 Correlation between selected countries' annual per capita chocolate consumption and the number of Nobel laureates per 10 million population

Source: Messerli 2012.

of Nobel Prizes awarded to individuals from different countries.

1 What conclusion do you draw from the graph in Figure 3.2?
2 Take a critical orientation to how the graph might be interpreted.
   - What other conclusions might you draw?
   - What are the anomalies represented in the graphical representation of the underlying data? (Hint: Look at the positions of countries you might expect to have more Nobel Prizes.)
3 Find other examples of data or graphical representations being used to create a deceiving impression.
4 Describe how you have been critical in responding to this activity.

THE NUMERACY MODEL IN ACTION

So far in this chapter we have focused on the individual dimensions of the numeracy model. In this section, two case studies are presented to illustrate how all aspects of the numeracy model are needed to examine problems set in real-world contexts. First, the method for reporting Australia's unemployment rate is described, and you are challenged to examine the way results are reported. Second, a newspaper article reporting changes to the dimensions of a popular chocolate bar is outlined for you to examine critically. Both case studies require the deployment of all dimensions of the numeracy model.

Case study 3.1 Australia's unemployment

The newspaper article transcribed below was published in *The Guardian* on 17 October 2017. It portrays a trend in relation to the issue of unemployment.

> **Australia's unemployment rate falls to four-year low of 5.5%**
> National Disability Insurance Scheme contributes to a surge in health jobs and Western Australia continuing to recover
>
> Australia's unemployment rate has fallen to a four-year low, with the labour market recording 12 straight months of employment gains—its longest stretch since 1994.
>
> Figures show the unemployment rate fell to 5.5% in September, its lowest rate since March 2013, as the government's National Disability Insurance Scheme contributes to a surge in health jobs and Western Australia continues to recover from the downturn in mining construction, adding 48,000 jobs over the last year.

While accepting that a decreasing unemployment rate is a very positive trend, it is worth considering what unemployment figures mean in a technical sense. The Australian Bureau of Statistics (2018a) calculates the unemployment rate as follows:

$$\text{unemployment rate} = \frac{\text{number employed}}{\text{labour force}} \times 100\%$$

where the labour force is the number of employed people plus the number of unemployed people. The Australian

Bureau of Statistics (2012) uses the following categories to collect information on people's employment status.

1. Want to work, actively looking for work, available to start work immediately
2. Want to work, actively looking for work, available to start work within 4 weeks
3. Want to work, actively looking for work, not available to start work within 4 weeks
4. Want to work, not actively looking because they believe they wouldn't be able to find a job, but would be able to start within 4 weeks
5. Want to work but not actively looking and not available to start within 4 weeks
6. Don't want to work
7. Permanently unable to work

Question

Discuss with a partner which of these categories you think represent individuals who are unemployed.

In fact, only people in Category 1 are counted as unemployed. Those in Categories 2, 3 and 4 are all classified as marginally attached to the labour force and so not considered as unemployed. Categories 5, 6 and 7 are considered to have no marginal attachment to the workforce and are also not counted as unemployed. (More information on these categories is available from the Australian Bureau of Statistics [2018b].)

While figures vary across a year, in February 2018 there were 12.5 million people employed, 722,000 unemployed and 6.7 million not in the workforce. Of those not considered in the workforce, 1.1 million were marginally attached (wanted to work but not actively looking for work, and available to start work last week or within four weeks) (Australian Bureau of Statistics 2018c).

Using the equation given above to calculate the unemployment rate reveals an unemployment rate of 5.5 per cent:

$$\frac{722,000}{13,222,000} \times 100\% = 5.5\%$$

Questions

1. Discuss with a partner why the figure in the equation is different from the one reported in the *Guardian* article.
2. By performing an appropriate calculation, find the difference between the previously calculated unemployment rate and the rate if those categorised as marginally attached were added to the workforce.
3. Comment on the significance of the 'new' unemployment rate.
4. Write a paragraph using the different dimensions of the 21st Century Numeracy Model to explain how a numerate person would interpret the ABS method of calculating unemployment rates.

Case study 3.2 Reducing the size of chocolate bars

The article transcribed below was published in *The Sunday Mail* (Queensland) on 30 May 2009.

Choc bars getting smaller

The Mars Bar—Australia's biggest-selling snack—will become 11 per cent smaller and it's all in the interests of our health, its maker claims.

The price, however, will remain the same.

Confectionery giant Mars Australia announced yesterday it has reduced the weight of its major chocolate bar brands.

Mars Bars and Snickers have dropped in size from 60g to 53g, making them 11.6 per cent lighter—and effectively more expensive.

King-size Mars Bars and Snickers drop from 80g to 72g. They are divided into two portions. Mars Snackfood Australia general manager Peter West said the move was a response to Australia's obesity debate because their chocolate bars had too many calories. 'Big food is an issue,' he said.

'The portion size that people are eating, whether the muffin you buy, or the cake, or when you go to a restaurant, the size of the plate in front of you—we want to make sure we sell our portions in the right size.'

Mr West said 30 million Mars Bars were sold in Australia every year. Based on that figure the company will save 210 tonnes of chocolate, caramel and nougat every year with the smaller bar.

Mr West said the current price of Mars Bars (from $1.70) had not fallen, although some of its other brands would be cheaper. All of Mars' 90 products will be downsized including the Milky Way, Snickers, Maltesers and Twix.

In advertisements today, Mars Snackfood says: 'Our products are never going to be as healthy as a piece of fruit.

'But we also recognise that our consumers have become increasingly concerned about the nutritional content and portion size they eat.'

The company says it will use clearer nutritional

labelling and remove artificial colours and flavours from Mars, Snickers and Milky Way.

Nutritionist and GP Manny Noakes welcomed the decision to slim down the bars, saying: 'It's a tremendous move. It's all about portion control and paying the same for less is not a bad thing and I think we can still enjoy it.'

Anti-obesity campaigner, federal MP Steve Georganas, said the company was 'just being cheeky'. 'If they're reducing the size by 11.6 per cent then they should reduce the price by 11.6 per cent,' he said. The Hindmarsh Labor MP will unveil a national obesity report tomorrow.

In the article above, the change to the size of the chocolate bars is portrayed as an ethical decision taken by the company in support of improving public health. The argument is that if the size of the bars is reduced (by some 11.6 per cent), people will eat smaller amounts of the less healthy components of this confectionary.

Questions

1. Comment on the justification Mars puts forward for reducing the size of its chocolate bars.
2. Make use of information in the article, and other data, to estimate how much money Mars will save each year due to the proposed change.
3. Do you think the change in size of the chocolate bars will lead to the proposed health benefits? What are other possible outcomes? Use a mathematical argument to support your position.
4. Discuss with a partner how you made use of each of the dimensions of the 21st Century Numeracy Model in responding to the questions related to this issue.

STRENGTHS AND LIMITATIONS OF THE MODEL

It is important to remember that the 21st Century Numeracy Model was originally developed to describe the nature of numeracy in the 21st century and to help teachers recognise numeracy demands and opportunities in different learning areas. Our work in schools has shown that the model can be effective in assisting teachers to create and structure numeracy tasks. The model has been used in this way to develop numeracy-focused activities in a range of learning areas—for example, English (Geiger, Goos, Dole et al. 2013), health and physical education (HPE; Peters et al. 2012), civics education (Willis et al. 2012) and environmental education (Cooper et al. 2012). These and other examples will be explored in Chapter 5.

The numeracy model has also been used to identify numeracy demands and opportunities in curriculum documents across learning areas (e.g., Goos, Dole et al. 2012). This provides starting points for teachers to think about where a focus on numeracy in a particular learning area might enhance the teaching of that learning area as well as promoting students' numeracy capabilities. How to identify numeracy demands and opportunities is addressed in Chapter 4 and Chapter 5, respectively.

While teachers have used the model to develop rich numeracy tasks, the model itself does not provide direction for how to promote numeracy capability—that is, how to help someone else to extend their numeracy capacities into broader or more sophisticated contexts. This requires guidance for designing numeracy tasks and how to plan for their effective implementation. The development of these important skills is now an expectation of the Australian Professional Standards for Teachers (AITSL 2017) under Standard 2: 'Know and understand literacy and numeracy teaching strategies and their application in teaching areas'. How teachers design and implement effective numeracy tasks has been another facet of our research (e.g., Geiger, Goos, Dole et al. 2014), which will be outlined and discussed in Chapter 6.

CONCLUSION

In this chapter, the 21st Century Numeracy Model (Goos, Geiger & Dole 2014) was introduced and illustrated via a range of real-world examples. These examples highlighted the role of mathematical knowledge, dispositions, tools and a critical orientation in solving problems in real-world contexts. The examples also served to demonstrate the importance of being numerate in personal, work and civic life. In Chapter 4, the model will serve as a focus for identifying the numeracy demands within the school curriculum.

RECOMMENDED READING

Attard, C., Ingram, N., Forgasz, H., Leder, G. & Grootenboer, P., 2016, 'Mathematics education and the affective domain', in K. Makar, S. Dole, J. Visnovska, M., Goos, A. Bennison & K. Fry (eds), *Research in Mathematics Education in Australasia, 2012–2015*, pp. 73–96, Singapore: Springer

Gapminder, 2018a, bubbles graph showing CO_2 emissions (per capita) over time, <http://bit.ly/2mFzmhN>, retrieved 25 July 2018

Gapminder, 2018b, bubbles graph showing cumulative CO_2 emissions over time, <http://bit.ly/2OfqN9W>, retrieved 25 July 2018

Geiger, V., Goos, M. & Dole, S., 2015, 'The role of digital technologies in numeracy teaching and learning', *International Journal of Science and Mathematics Education*, vol. 13, no. 5, pp. 1115–37, doi: 10.1007/s10763-014-9530-4

Goos, M., Geiger, V. & Dole, S., 2014, 'Transforming professional practice in numeracy teaching', in Y. Li, E. Silver & S. Li (eds), *Transforming Mathematics Instruction: Multiple approaches and practices*, pp. 81–102, New York: Springer

Quantitative Literacy Design Team, 2001, 'The case for quantitative literacy', in L. Steen (ed.), *Mathematics and Democracy: The case for quantitative literacy*, pp. 1–22, Princeton, NJ: National Council on Education and the Disciplines

Rosa, M. & Clark Orey, D., 2015, 'A trivium curriculum for mathematics based on literacy, matheracy, and technoracy: An ethnomathematical perspective', *ZDM Mathematics Education*, vol. 47, no. 4, pp. 587–98

4

Numeracy demands

Developing numeracy requires that students gain the confidence and experience to use their mathematical knowledge, not only in everyday situations but also in all the subjects they study at school. This chapter shows how to use the 21st Century Numeracy Model introduced in Chapter 3, to make visible the numeracy demands of different learning areas in the school curriculum. To understand what numeracy across the curriculum looks like, we distinguish between numeracy *demands* and numeracy *opportunities*. We argue that every learning area—such as history, geography or science—has its own inherent numeracy demands, which can be identified by analysing its published curriculum. We refer to the results of such an analysis as the numeracy *fingerprint* of the learning area. Teachers can also create different kinds of numeracy learning opportunities through the classroom strategies they implement, and this is a topic addressed in Chapter 5.

The first section of this chapter investigates the nature of numeracy as a general capability in the Australian Curriculum and evaluates the tools embedded in the curriculum to help teachers see how numeracy is used or developed in different learning areas. We then demonstrate our alternative approach to revealing the curriculum's numeracy demands by looking through the lens of our numeracy model. Finally, we consider the roles of mathematics and mathematics teachers in developing cross-curricular numeracy.

NUMERACY IN THE AUSTRALIAN CURRICULUM

Numeracy is one of seven general capabilities in the Australian Curriculum that teachers are to address through the content of the learning areas. The Australian Curriculum describes numeracy in the following terms: 'Numeracy encompasses the knowledge, skills, behaviours and dispositions that students need to use mathematics in a wide range of situations. It involves students recognising and understanding the role of mathematics in the world and having the dispositions and capacities to use mathematical knowledge and skills purposefully' (ACARA 2018e).

The Australian Curriculum makes it clear that all teachers are responsible for developing their students' numeracy:

> This means that all teachers:
> - identify the specific numeracy demands of their learning area/s;
> - provide learning experiences and opportunities that support

> **Review and reflect 4.1**
>
> 1 Visit the 'Numeracy' section of the Australian Curriculum website (ACARA 2018e) and explore the organising elements for numeracy outlined there: 'Estimating and calculating with whole numbers'; 'Recognising and using patterns and relationships'; 'Using fractions, decimals, percentages, ratios and rates'; 'Using spatial reasoning'; 'Interpreting statistical information'; and 'Using measurement'.
> 2 Compare the elements of numeracy as proposed by the Australian Curriculum with the elements of the 21st Century Numeracy Model introduced in Chapter 3.
> 3 Discuss the similarities and differences with a partner.

the application of students' general mathematical knowledge and skills;

- should be aware of the correct use of mathematical terminology in their learning area/s and use this language in their teaching as appropriate. (ACARA 2018e)

These expectations place significant demands on teachers at all levels of schooling and in all learning areas. Two kinds of tools—icons and filters—are embedded in the online Australian Curriculum to help teachers meet these expectations.

Numeracy icons and filters

The Australian Curriculum website uses icons to indicate where general capabilities have been identified within the learning area content descriptions and elaborations. The icons are visible in the online curriculum for all learning areas, and a filter can be applied to identify numeracy demands within this content.

As an example of the use of the numeracy filter, consider the learning area of HPE in Years 7 and 8 (ACARA, 2018c). The curriculum filter can be set to include the Years 7 and 8 band, together with the general capabilities icon for numeracy. Clicking 'Submit' with these settings selected returns the result 'No content at this level'; in other words, the filter does not identify any numeracy demands in the Year 7 and 8 HPE curriculum. This seems to be a surprising result when the curriculum at these year levels includes focus areas such as food and nutrition, health benefits of physical activity, and games and sports—all of which have inherent numeracy demands.

Table 4.1 lists three examples of content descriptions and associated content elaborations for Years 7 and 8 HPE, together with an indication of possible numeracy demands. None of these content descriptions is tagged with the numeracy icon, and so none is identified by the process of applying the online numeracy filter. However, all the elaborations are tagged with the icon to indicate they involve numeracy. Thus, while the online filtering tool is a quick and convenient way to identify numeracy within content descriptions, it does not reveal all the inherent numeracy demands captured by the content elaborations.

AUDITING THE NUMERACY DEMANDS OF THE SCHOOL CURRICULUM

Review and reflect 4.1 drew attention to the similarities and differences between the description of numeracy in the Australian

Table 4.1 Examples of numeracy demands in Years 7 and 8 HPE

Content descriptions	Content elaborations	Possible numeracy demands
Investigate and select strategies to promote health, safety and wellbeing	Research a variety of snack and lunch options, and evaluate nutritional value, value for money and sustainability impacts to create a weekly menu plan	• Interpret tables of nutritional composition of foods • Calculate nutritional value of specific foods in different portion sizes • Calculate cost per portion and cost per nutrient component
Participate in physical activities that develop health-and skill-related fitness components, and create and monitor personal fitness plans	Measure heart rate, breathing rate and ability to talk in order to monitor the body's reaction to a range of physical activities, and predicting the benefits of each activity on health- and skill-related fitness components	• Use a watch to measure time while counting heartbeats and breaths • Calculate heart and breathing rates • Construct a table or graph to compare heart and breathing rates in response to different activities
Demonstrate and explain how the elements of effort, space, time, objects and people can enhance movement sequences	Demonstrate an understanding of how to adjust the angle of release of an object and how this will affect the height and distance of flight	• Use appropriate tools to measure angle of release, height, distance • Construct a table or graph to compare effect of angle on height and distance

Source: ACARA 2018c (for content descriptions and elaborations).

> **Review and reflect 4.2**
>
> 1 Review the 'Numeracy in the learning areas' statement for the Australian Curriculum in HPE (ACARA 2018e). Compare the information in this statement with the numeracy demands suggested in Table 4.1.
> 2 With a small group of partners (in different learning areas if secondary teachers), investigate the use of numeracy icons and filters for two different learning areas in the Australian Curriculum, within a chosen year level.
>   - Record the content descriptions identified by the numeracy filter for your chosen year level.
>   - Reset the filter so that all the content descriptions are visible and check for elaborations that are tagged with the numeracy icon but not captured by the filtering process (as in Table 4.1).
> 3 Discuss the possible nature of the numeracy demands associated with each of these elaborations.
> 4 Compare your findings with the information contained in the relevant 'Numeracy in the learning areas' statement in the Australian Curriculum for these learning areas.

Curriculum and the 21st Century Numeracy Model introduced in Chapter 3. For example, our numeracy model emphasises the use of tools as well as mathematical knowledge and skills, places contexts at the centre of numeracy and highlights the significance of a critical orientation to the world. One of the ways in which the numeracy model has been used is to audit the distinctive numeracy demands of the different learning areas in the school curriculum, which we refer to as the learning areas' numeracy fingerprints. The audit strategy was developed by the authors of this book before the Australian Curriculum was introduced, and the approach can be applied to any curriculum document.

We will illustrate our approach by drawing on the numeracy audit that we conducted of the former South Australian Curriculum, Standards and Accountability (SACSA) Framework (Department of Education and Children's Services 2005). There are two reasons for demonstrating the numeracy audit approach on an archived curriculum document rather than on the current Australian Curriculum. First, we want to establish that the audit process can be applied to any curriculum; and second, we leave it to you, the reader, to conduct your own numeracy audit of the learning areas you teach in the Australian Curriculum.

A summary of our full numeracy audit findings is available in a booklet produced by the South Australian Department of Education and Children's Services (2009). The audit evaluated the numeracy demands of the arts, design and technology, English, HPE, languages, mathematics, science, and society and environment learning areas for Years 6 to 9, as represented by the relevant curriculum scope and standards statements provided in the SACSA Framework.

Numeracy demands of each learning area were evaluated by reference to the elements of the numeracy model introduced in Chapter 3. Mathematical knowledge demands were examined by assessing the extent to which the target learning area drew on the five strands of the mathematics learning area of the SACSA Framework:

- 'Exploring, analysing and modelling data' involves collecting, organising, representing, analysing and interpreting data from real-world contexts and applying ideas about chance and probability.
- 'Measurement' involves understanding attributes, units and systems of measurement in a range of contexts and solving problems involving measurement.
- 'Number' involves using number concepts and operations to make sense of real-world activities and using appropriate computational methods to solve problems.
- 'Patterns and algebraic reasoning' involves recognising patterns in real-world contexts, making generalisations and investigating change and using algebraic formulae to solve problems.
- 'Spatial sense and geometric reasoning' involves exploring the environment in terms of spatial sense, location and movement and producing maps and graphs (Department of Education and Children's Services 2005).

Auditing numeracy demands in the arts

The following sample evaluation of the arts learning area models our approach to auditing the numeracy demands of the whole curriculum. The SACSA Framework's statement about how students develop their operational skills in numeracy in the arts (which is similar to statements about numeracy as a general capability in the different learning areas in the Australian Curriculum) is quoted below:

> Learners develop and use operational skills in *numeracy* to understand, analyse, critically respond to and use mathematics

in different contexts. These understandings relate to measurement, spatial sense, patterns and algebra and data and number. This learning is evident in arts when, for example, students design products using sequencing and patterning, accurate measurement and a sense of shape, size, dimension and perspective. Gathering, interpreting and analysing data in relation to audience, viewer and user behaviour is another example of numeracy in arts. (Department of Education and Children's Services 2005, p. 13, original emphasis)

The framework also lists the strands and key ideas that organised the curriculum scope for this learning area.

Strand: arts practice
Key idea[s]
Students draw from thought, imagination, data and research, and the examination of social and cultural issues, to demonstrate personal aesthetic preference, and provide imaginative solutions and artistic responses to ideas and issues. (Department of Education and Children's Services 2005, p. 16)

Students develop knowledge of the styles, forms and conventions of each arts form; refine arts skills; apply appropriate techniques; explore, plan, organise and employ both creative and abstract thought in the production of arts works. (p. 18)

Students develop their capacity to interact effectively with people from a diversity of interests and abilities. They learn to work as individuals and as members of production/performance teams and to assume specific roles and responsibilities in the development and production of arts works which achieve particular responses from audiences/viewers. (p. 22)

Strand: arts analysis and response
Key idea
Students learn to distinguish different genres and styles associated with the different arts forms. They employ processes for analysis and interpretation of style, genre and form of arts works, and communicate both reasoned and personal viewpoints in response to arts works. (p. 26)

Strand: arts in contexts
Key idea[s]
Students examine and analyse their knowledge of a wide range of arts works, the arts industry and social influences to understand the impact of these on their own and their peers' work and that of Australian contemporary artists. (p. 30)

Students investigate the arts practices of a number of cultures across time to develop an understanding and appreciation of the cultural and global connections which are emerging as a result of social and technological change. (p. 32)

Mathematical knowledge

The SACSA Framework for mathematics identified five strands, as listed above. Table 4.2 maps numeracy learning demands in the arts onto these mathematics strands.

Learning in the arts makes many numeracy demands of students in terms of mathematical knowledge. In arts practice, where students

Table 4.2 Mathematical knowledge demands within strands of the arts learning area of the SACSA Framework

Mathematics strands	Arts strands		
	Practice	Analysis/response	In contexts
Exploring, analysing and modelling data			
Measurement			
Number			
Patterns and algebraic reasoning			
Spatial sense and geometric reasoning			

Note: Shading is used to indicate moderate (light shading) and high (dark shading) levels of numeracy learning demands.

create arts works, they draw on elements of the mathematics strands in exploring and analysing data and using measurement, number and spatial sense as well as patterns (which supports algebra understanding but may not promote algebra understanding specifically). In arts analysis and response, where students analyse others' arts works, they may conduct research that requires analysis of data and use spatial reasoning to consider elements of arts works.

Contexts

The SACSA Framework explains that the arts give voice to thoughts, feelings and beliefs through five main art forms: 'dance, drama, media, music and visual arts ... All styles of expression described by such terms as traditional, contemporary, popular, folk, commercial and fine arts are represented in arts works. Individual arts works can serve to maintain the status quo or challenge assumptions, and critique social, cultural, economic and political practices' (Department of Education and Children's Services 2005, p. 9).

The arts provide contexts for understanding symbols and symbolism and to symbolise moments of great importance. Arts can be linked to mathematics through the study of signs and symbolism. In arts practice, students design products and use mathematics to consider sequencing and patterning, shape, size, dimension and perspective. They gather, interpret and analyse data in relation to audience, viewer and user behaviours. In visual arts, students use mathematics to consider colour creation, texture and emotion. The mixing of paints requires use of proportional reasoning through exploration and analysis of amounts of various colours to create different tints and hues. Composing dance, drama and visual art involves the mathematical concepts of symmetry, proportion, reflection, rotation and translation.

When they use a newspaper as a stimulus for exploring an issue, students need to use mathematics to analyse and go beyond facts and information provided. When students engage in performances, exhibitions, festivals and cultural events, they need mathematics to project costings, to analyse space and built environment restrictions and affordances, to explore timelines and to plan sequences of events. When considering paid and unpaid arts works, they are using critical numeracy skills to evaluate the financial costs of arts events and works with the social, emotional and spiritual benefits of arts in the community. When playing an instrument, creating songs and dances, they are using mathematics to consider beat and rhythm and the mathematical structure of music and melody.

To make a social comment through drama, mathematics is used to enhance understanding of the issue through research (e.g., what would happen if all Australians gave $5 each week to a charity?). Through critical analysis of the media, students use mathematics to support examination of codes and conventions. In designing and critiquing arts works, students use mathematics to consider aesthetics of design.

Dispositions

Major dispositions promoted through the arts curriculum area are perseverance and commitment, with students planning and creating arts works, seeing arts works through from earliest stages to completion. While perseverance and commitment can also support students as they engage in mathematical problem-solving investigations, positive dispositions in one learning area do not necessarily transfer to another part of the curriculum. Teachers need to be alert to the possibility that students who enjoy and persist in working in the arts might not be developing positive dispositions to numeracy.

Tools

Students use a range of representational, physical and digital tools in creating and responding to arts works. In music, for example, they make notations to symbolise rhythm, melody, harmony and tempo. In the visual arts, they work with materials, implements, images and media to communicate and represent ideas in two and three dimensions. In drama, they make models of performance spaces when they study set design and construction and consider the use of space to suit a scene from a play. They become familiar with print, film and electronic media as well as web-based resources such as virtual galleries and libraries, and they use many types of digital technologies to create new art forms.

Critical orientation

The arts have the potential to encourage students to move beyond arts works for self-expression to arts works for social comment and critique. In this way, the arts can be the vehicle through which students develop a critical orientation, where they use mathematics tools to analyse facts and information about situations and issues. For example, students might investigate how images and colours are manipulated to design logos or advertising materials. They can then create their own images, using the same techniques, that aim to persuade viewers to buy a product or support a business or sporting team.

Numeracy fingerprints in the school curriculum

Our audit of the South Australian curriculum for Years 6 to 9 revealed the distinctive numeracy fingerprints of each of the learning areas.

Mathematical knowledge

Table 4.3 synthesises the mathematical knowledge requirements of the seven learning areas in the SACSA Framework, apart from mathematics itself. (We consider the role of mathematics in developing students' numeracy later in this chapter.) The synthesis was carried out by combining the mappings of numeracy learning demands of each strand in each learning area onto the mathematics strands—for example, as depicted in Table 4.2. For each of these mappings, scores of 2 were allocated for strands with high numeracy demands, 1 for strands with moderate numeracy demands and 0 for strands with low numeracy demands. These scores were then tallied, by mathematics strand, for each learning area. This procedure resulted in each cell of Table 4.3 representing a score between 0 and 6 (three strands per learning area, each with a minimum score of 0 and a maximum score of 2).

From Table 4.3, we see that the level of numeracy demand was highest for design and technology, science and the arts; moderate for society and environment and HPE; and lowest for English and languages. Despite these differences, however, it is important to recognise that *all* learning areas had distinctive numeracy

Table 4.3 Mathematical knowledge demands within learning areas of the SACSA Framework (excluding mathematics)

Mathematics strands	Learning areas						
	Arts	DT	Eng	HPE	Lan	Sci	SE
Exploring, analysing and modelling data							
Measurement							
Number							
Patterns and algebraic reasoning							
Spatial sense and geometric reasoning							

Note: DT design and technology; Eng English; HPE health and physical education; Lan languages; Sci science; SE society and environment. Shading is used to indicate the level of numeracy demand: low (or a score of 0–1) is unshaded, moderate (2–4) has light shading, and high (5–6) has dark shading.

Source: Goos, Geiger & Dole (2010).

demands in relation to the type of mathematical knowledge required by students in order to demonstrate successful learning. Teachers are ultimately responsible for enacting the curriculum in their classrooms, and they can therefore exploit numeracy learning opportunities beyond those implied by the published curriculum.

The strands of mathematical knowledge were also represented to different extents in the SACSA Framework's learning areas. 'Exploring, analysing and modelling data' was most strongly represented in the intended curriculum, followed by 'Measurement', 'Number' and 'Spatial sense and geometric reasoning', with the strand of 'Patterns and algebraic reasoning' the least strongly represented. It is perhaps not surprising that algebra, as an element of numeracy knowledge, appeared to be underrepresented in the curriculum, since it is often considered to be quite abstract, with little connection to real-world contexts or learning areas other than mathematics. However, it is worth emphasising the potential connection between algebraic reasoning and modelling with data, since exploration of patterns and generality in the middle years of schooling can begin with an empirical focus on data collection and analysis.

Contexts

The range of numeracy contexts highlighted in the 21st Century Numeracy Model was well represented across the learning areas of the SACSA Framework. Each learning area emphasised the value of connecting students' learning to real-life contexts that are meaningful for them, whether this involves personal interests, family and community life, leisure pursuits, the physical environment, vocations and careers, diverse cultures or social, economic and political systems.

Dispositions

Throughout the SACSA Framework there was evidence of a desire to develop positive dispositions, such as perseverance, confidence, resilience, willingness to take risks and show initiative, respect for cultural diversity and commitment to ecological sustainability. These are admirable goals, but we would point out that dispositions towards learning in one discipline do not automatically transfer to another discipline: it is possible, for example, for students to feel confident about their learning in the arts but not in mathematics and not in relation to numeracy more generally. Teachers need to be aware of the damaging effects of negative mathematical dispositions, to look for opportunities to successfully engage their students with the numeracy demands of their learning area and to make explicit to students the positive dispositions that are helping them to achieve this success. Linking positive dispositions with numeracy learning is therefore vital.

Tools

Representational, physical and digital tools were used across all learning areas in the SACSA Framework. Some of these were specific to the discipline, while others were more generically useful. Graphs, diagrams, tables, maps and plans were commonly used in many learning areas, as were measuring instruments, both physical and digital. There was also a strong emphasis on digital tools, software and web resources. Thus, all learning areas had specific numeracy demands in relation to accurate and intelligent use of tools to represent and analyse ideas. Students need to become proficient with the tools of each learning area, but they also need to be aware that some tools are used in more than one learning area and to be flexible in applying tools in different curricular contexts. For example, students may come to believe that there are different ways to read and create maps in mathematics and in society and environment or to create graphs that show relationships between variables in science and in HPE. Teachers in these learning areas need to be aware of any differences in techniques and terminologies associated with the use of these representational tools and to draw students' attention to important similarities between underlying concepts.

Critical orientation

The SACSA Framework emphasised developing a critical orientation in students across all learning areas. Such an orientation could not be fully enabled without numeracy knowledge, dispositions and tools, nor could it be convincingly enacted unless learning took place in a range of real-life contexts. Conversely, being numerate requires adopting a critical stance in order to question, compare, analyse and consider alternatives.

MATHEMATICS AND CROSS-CURRICULAR NUMERACY

Embedding numeracy across the school curriculum does not require that all teachers become specialist teachers of mathematics. And as Kemp and Hogan (2000, p. 23) argued, 'Approaching a situation from the point of view of teaching some mathematics is not necessarily the same as a student using some mathematics to help them understand or do something else'. The purpose of understanding the numeracy demands of the subject you are teaching is to enhance students' learning of this subject, not to teach them more mathematics.

> **Review and reflect 4.3**
>
> For this task, work with a partner who has a specialisation or interest in the same learning area as you.
>
> 1 Read the booklet summarising the findings of the SACSA middle years numeracy audit (Department of Education and Children's Services 2009). Also, read the following three short conference papers, which provide further examples of numeracy audits of the curriculum: Goos, Geiger & Dole 2010 (a sample audit of the society and environment curriculum); Goos, Dole et al. 2012 (a sample audit of the history curriculum); Geiger, Goos, Dole et al. 2013 (a sample audit of the English curriculum).
> 2 With your partner, choose one year level in the Australian Curriculum for your preferred learning area and conduct a numeracy audit of this year level using the 21st Century Numeracy Model as modelled in this chapter. In addition to the content descriptions and elaborations, you will need to read the learning area aims and rationale and to locate the advice on numeracy specific to the learning area.
> 3 Present your findings to another pair of colleagues who have conducted a numeracy audit of a different learning area. Compare the numeracy fingerprints of the two learning areas.

Teachers of mathematics in primary and secondary schools nevertheless have a responsibility to ensure that students are learning mathematics well and developing mathematical knowledge and skills that may then be applied in other curriculum contexts (Hogan 2000a). Another important role of the specialist mathematics teacher, whether this is in the primary or the secondary school, is to act as a resource person for colleagues who are teaching in other learning areas, to assist with making numeracy visible and to collaborate on designing tasks that develop students' numeracy within those areas.

Auditing numeracy demands in mathematics

Previously, we illustrated our approach to auditing the numeracy demands of the school curriculum—but what of the numeracy demands of mathematics itself? We can apply a similar audit approach, which we again illustrate by referring to the former SACSA Framework for Years 6 to 9.

Mathematical knowledge

From a numeracy perspective, effective mathematics teaching demands that students see connections between the different mathematical content strands and between mathematics and the real world. While this seems self-evident, it is easy to overlook chances to explore ideas in 'Measurement', for example, while working within the strand of 'Patterns and algebraic reasoning'. Knowledge of and capacity to make use of the interconnectedness of mathematical ideas, represented in the curriculum content strands, are vital elements in being effectively numerate. Investigations into life-related contexts are rich in these types of connections. For example, a student might measure the height of a plant grown from a seed over a series of weeks and then graph these data. They might also attempt to find a pattern in the rate of growth based on both numeric and graphing data. The investigation might be extended to investigate the effects on growth of different planting arrangements of the seedlings in a garden plot. Each of these examples highlights the numeracy demands of the mathematics learning area in terms of developing mathematical knowledge.

Contexts

The learning area of mathematics in the SACSA Framework aimed to empower students, through the capacity to use mathematical ideas critically, to become active and constructive citizens of their society. This is a broader role for mathematics than that of the study of a subject for its own sake, and more expansive than that of an enabling discipline for other learning areas. Numerate citizens must use mathematics in order to explore the cultural and social issues of the world as well as the scientific and economic changes that are taking place in society.

Dispositions

Students' dispositions towards mathematics are inevitably influenced by their experiences in mathematics lessons. By the time they reach the later primary years, many students have come to believe that they are not successful at mathematics, that learning mathematics involves memorising facts and formulae or that there is always one correct way of solving any problem. These beliefs are not conducive to developing positive numeracy dispositions, such as confidence, flexibility, initiative and willingness to take intellectual risks. Emphasising applications of mathematics has a vital role to play within the development of numeracy. To apply mathematics to the real world requires students to develop as flexible thinkers with the confidence to show initiative and a disposition towards taking calculated risks—to trial an approach, to evaluate the success or otherwise of an attempt and then to make the necessary changes to improve their response to a task.

Tools

The use of tools in mathematics has a long history, especially in areas that are related to calculation and geometry. The role of digital tools in learning and teaching mathematics has also become increasingly important. Tools were used in the SACSA Framework's mathematics learning area in the following ways:

- rulers, measuring tapes, trundle wheels and a variety of survey equipment used to determine distances, altitudes and relevant positions of landmarks
- geometrical tools used for determining comparative distances and angles between objects or landmarks on scaled maps
- data-logging devices that collect information in real time on life-related phenomena, such as velocity, distance, gas absorption and applied force
- online data sources, such as archival records of athletic performance
- digital tools, such as software applications and learning objects developed for the exploration of real-world phenomena.

Critical orientation

Mathematics can provide vital analysis tools for the critical examination of claims made or opinions expressed by politicians and policy-makers. These tools are also important for citizens to understand and engage with the processes of cultural, social and environmental changes that are taking place globally. It is important within mathematics that an outward orientation towards other real-world and curriculum contexts is developed, so that such issues are brought into the mathematics classroom.

> ### Review and reflect 4.4
>
> Work in pairs comprising a mathematics specialist and a teacher of another learning area.
>
> 1. Select a unit plan from this term's work in a non-mathematics subject.
> 2. Identify a segment of this plan that is potentially numeracy rich and discuss its numeracy demands.
> 3. Prepare one lesson for the non-mathematics subject that stimulates students' numeracy learning in that subject and one mathematics lesson that develops some of the relevant mathematical concepts that students will need to apply in their learning of the non-mathematics subject.

CONCLUSION

In this chapter, we have introduced an important distinction between the numeracy demands of the curriculum, which can be identified through analysis of the published curriculum documents, and numeracy opportunities, which teachers can create through the strategies they implement in their planning and classroom practice. We have also modelled an approach to auditing the numeracy demands of all learning areas across the school curriculum, drawing on an audit we conducted of a state-based curriculum before the introduction of the Australian Curriculum. Our audit approach involves applying the elements of the 21st Century Numeracy Model—mathematical knowledge, contexts, dispositions, tools and a critical orientation—to examine relevant sections of the curriculum document. This approach can be used for any published curriculum to reveal its inherent numeracy demands—that is, its numeracy fingerprint.

RECOMMENDED READING

Department of Education and Children's Services, 2009, *Numeracy in the Middle Years Curriculum: A resource paper; An audit of numeracy in the SACSA framework*, <https://numeracy4schools.files.wordpress.com/2015/03/numeracy-audit-book_v6.pdf>, retrieved 12 March 2018

Geiger, V., Goos, M., Dole, S., Forgasz & Bennison, A., 2013, 'Exploring the demands and opportunities for numeracy in the Australian Curriculum: English', in V. Steinle, L. Ball & C. Bardini (eds), *Mathematics Education: Yesterday, today and tomorrow*, Proceedings of the 36th annual conference of the Mathematics Education Research Group of Australasia, vol. 1, pp. 330–7, Melbourne: MERGA

Goos, M., Dole, S. & Geiger, V., 2012, 'Auditing the numeracy demands of the Australian Curriculum', in J. Dindyal, L. Chen & S.F. Ng (eds), *Mathematics Education: Expanding horizons*, Proceedings of the 35th annual conference of the Mathematics Education Research Group of Australasia, pp. 314–21, Singapore: MERGA

Goos, M., Geiger, V. & Dole, S., 2010, 'Auditing the numeracy demands of the middle years curriculum', in L. Sparrow, B. Kissane & C. Hurst (eds), *Shaping the Future of Mathematics Education*, Proceedings of the 33rd annual conference of the Mathematics Education Research Group of Australasia, pp. 210–17, Fremantle: MERGA

5

Numeracy opportunities

One avenue for developing students' numeracy capabilities is to identify numeracy demands in published curriculum documents. In Chapter 4, we saw how to use the 21st Century Numeracy Model introduced in Chapter 3, to identify inherent numeracy demands of different learning areas in the school curriculum. We drew a distinction between these demands and numeracy opportunities. Numeracy opportunities are evident in published curriculum documents for all learning areas but are invisible unless one knows how to 'see' them (Goos, Dole et al. 2012). This chapter illustrates how to examine the curriculum in different learning areas to identify numeracy opportunities. Once teachers identify numeracy demands and opportunities, they need to plan for numeracy learning by designing tasks that enhance both numeracy and subject learning. Chapter 6 addresses principles for achieving this objective.

The first section of the chapter investigates the sometimes hidden role of numeracy in learning of disciplinary content in tertiary courses taken by prospective teachers. Next, we provide examples where teachers have recognised numeracy opportunities not identified in curriculum documents. We then demonstrate a process for identifying numeracy opportunities in the published curriculum. Finally, we consider aspects of numeracy that are inherent in several learning areas across the curriculum.

DISCIPLINARY CONTENT KNOWLEDGE

Teachers need to 'know the content and how to teach it' (AITSL 2017). According to Shulman (1987, p. 8–9), content knowledge encompasses 'the knowledge, understanding, skill, and disposition that are to be learned'. For this reason, initial teacher education programs at undergraduate level include courses that assist pre-service teachers to develop disciplinary content knowledge and at postgraduate level require them to have completed specific prerequisites that develop content knowledge for the subjects they will teach. Undergraduate teacher education programs include courses that address disciplinary content and pedagogy. Programs at postgraduate level focus on pedagogy, with less attention given to disciplinary content. Entrants to secondary programs must have at least a major study in one curriculum area, and entrants to primary programs must have at least one year's full-time-equivalent study in an area relevant to one or more learning areas in the primary curriculum (AITSL 2015, p. 12).

University courses undertaken by pre-service teachers to develop their disciplinary content knowledge do not usually address the role numeracy plays in disciplinary understanding. This is one reason why teachers can sometimes view numeracy as something additional to be included in an already crowded curriculum. However, it is possible to identify where having the capacity and disposition to engage with inherent mathematical knowledge in a disciplinary context not only enhances understanding of the discipline but also is essential. For example, having the capacity to read and interpret information presented in graphical form is necessary in many disciplines.

Review and reflect 5.1

Reflect on the courses in your university studies in which you developed the disciplinary content knowledge for the subjects you teach. If you teach in a primary school, consider one learning area other than mathematics.

1. Compile a list of instances where you needed to draw on your mathematical knowledge.
2. Rate these mathematical ideas as (a) enhancing your understanding of the discipline, (b) essential to learning the discipline or (c) both.
3. How did your mathematical knowledge contribute to your disciplinary learning? What challenges did you face?
4. Compare your ideas with a partner. Discuss ways in which explicit attention to numeracy might enhance learning, is essential in this learning area, or both. What challenges might your students encounter?
5. Share your ideas with another pair of colleagues who teach a different subject. Identify similarities and differences.

NUMERACY ACROSS THE CURRICULUM

The purpose of this chapter is not to provide a suite of ready-made activities that provide explicit links between numeracy and learning areas but to encourage you to think about how activities based on the examples presented could be applied to the subjects you teach. In the previous chapter, we saw how to identify numeracy demands within published curriculum documents and raised the question of how teachers could develop the capacity to see numeracy opportunities. One way of developing this capacity is to analyse activities in several different learning areas that draw on some mathematical knowledge. There are two questions to be answered in this analysis: How does the activity address curriculum goals in the target learning area? and How does the activity develop students' numeracy capabilities? We can answer the first question by examining activities through the lens of the relevant curriculum documents. For example, budgeting in history can help develop empathy; a scaled timeline in science assists understanding of geological time; and a problem provides a context in mathematics (see Table 5.1). For the second question in the analysis, examining activities through a numeracy lens using the 21st Century Numeracy Model provides insights into the dimensions of numeracy addressed by an activity.

Your capacity to see numeracy in the subjects you teach will increase as you become more experienced in looking for numeracy. For example, one teacher who worked with us over a three-year period commented that numeracy had become 'so much more, not prevalent, but obvious in what I do'. Once you are able to identify numeracy in the subjects you teach, you will be able to help your students to make explicit connections between their mathematical

Table 5.1 Examples of how numeracy can be embedded across the curriculum

Curriculum areas and topics	Descriptions of tasks	Curriculum goals
History: Australia in 1900	Students use information about weekly wages and the cost of daily requirements, such as food and transport, to prepare a weekly budget	To understand what life was like in Australia at the beginning of the 20th century
Science: Geological time	Students construct a geological time scale using a roll of paper towel and a measuring tape	To understand the extent of geological time (i.e., how little time plant and animal life has existed when looking at the history of the earth)
Mathematics: Measurement	Students determine the cost of floor tiles needed for the classroom and whether it is feasible to transport these from the place of purchase to school by car	To develop skills in using measuring tools and problem-solving

Source: Bennison 2015.

knowledge and how they can use this knowledge outside the mathematics classroom.

In the following section, we explore some activities in different learning areas to see how they address both curriculum goals within the relevant learning area and numeracy. The case studies come from a series of research and development projects in which we have used the 21st Century Numeracy Model to help teachers to identify and plan for numeracy in subjects across the curriculum.

HPE

The Australian Curriculum: Health and Physical Education includes two interrelated strands: 'Personal, social, and community health' and 'Movement and physical activity'. Each strand consists of three sub-strands. An excerpt from the content descriptor for the 'Understanding movement' sub-strand for Years 7 and 8 is shown in Figure 5.1 along with the relevant elaboration and elements of numeracy identified by the numeracy icon. Although the curriculum filter identifies a numeracy demand for this elaboration, Case study 5.1 illustrates potential for greater attention to numeracy—in other words, how the teacher created a numeracy opportunity. The case study also highlights the role digital tools can play in promoting the development of students' numeracy capabilities. (For further discussion of the role of digital tools to promote numeracy learning, see Geiger, Goos & Dole 2015.)

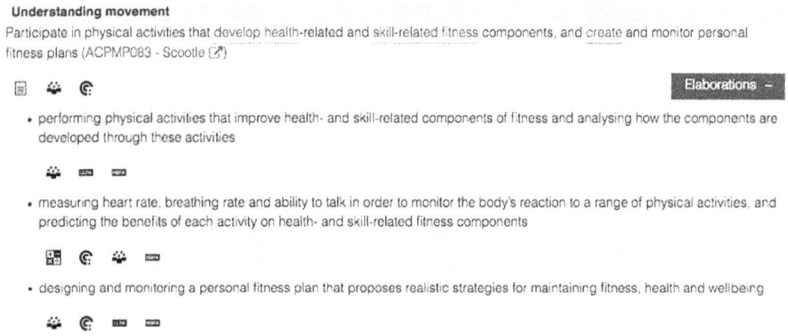

Figure 5.1 Excerpt from the 'Understanding movement' sub-strand of the Australian Curriculum: Health and Physical Education for Years 7 and 8

Source: ACARA 2018c.

Case study 5.1 Using a pedometer

Students in a Year 8 physical education class investigated their level of physical activity using a pedometer. Each student recorded in a shared Excel spreadsheet the number of paces taken each day for a week. Analysis of individual and class data for the week included using the graphing tool in Excel to produce different graphical representations of the data. Students also converted the total number of paces taken to the distance travelled, by working out how many paces they took to walk 100 metres, multiplying their answer by ten to find the number of paces to walk one kilometre, and dividing their total number of paces by this number. The teacher took an inquiry approach throughout the activity, using prompts to guide students rather than answering their questions. (For further details of this activity, see Peters et al. 2012.)

This activity addresses the HPE curriculum goal of having students participate in physical activities that develop health-related components of fitness. The activity could serve as a precursor to designing and monitoring a personal fitness plan. From a numeracy perspective, the activity takes place in the context of determining the level of physical activity and requires use of mathematical knowledge (measurement and ratio, selecting an appropriate graphical representation for the data) and a range of tools (pedometer, tape measures, calculators and Excel spreadsheet). Employing an inquiry approach to foster confidence and using digital tools in an outdoor activity to enhance student engagement with mathematics address the dispositions dimension of numeracy. Finally, making judgements about the reasonableness of their results and the appropriateness of particular graphical representations and speculating on possible reasons for differences between their results and those of others require students to take a critical orientation.

English

The Australian Curriculum: English includes three interrelated strands: 'Language', 'Literature' and 'Literacy'. From Year 2, students engage with a variety of texts, including those in which the primary purpose is aesthetic (including to entertain) and texts designed to inform and persuade. The types of texts that students engage with vary across year levels and in Year 7 include media texts such as newspapers. The content descriptors in the 'Creating texts' sub-strand of the 'Literacy' strand are shown in Figure 5.2. The numeracy icon does not appear, suggesting there are no inherent numeracy demands in any of this content. However, Case study 5.2 shows how a teacher created a numeracy opportunity for students learning this content. Specifically, it illustrates how numeracy can be incorporated into persuasive texts to support a chosen position or to refute contrary positions. This is particularly relevant for students in developing the critical aspects of numeracy. (For further discussion of the critical orientation dimension of numeracy, see Geiger, Forgasz et al. 2015.)

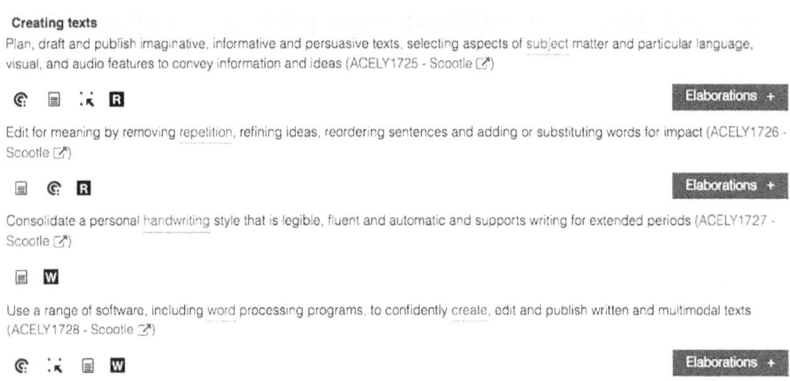

Figure 5.2 Excerpt from the 'Creating texts' sub-strand of the Australian Curriculum: English for Year 7

Source: ACARA 2018b.

Case study 5.2 Writing an opinion piece

Students in a Year 7 English class took turns to read an article retrieved from a newspaper's website about a proposed new law that included a clause that would impose a ten-year ban from owning a dog on owners of dogs that attacked people. The article included statistical information on recent dog attacks, including the number of attacks per year from 2002 until 2013, the age of people attacked and the parts of the body injured. Students discussed the article with particular attention to the data presented and associated interpretations. A second article was treated in a similar manner. Students were then asked to locate three additional articles on dog attacks that also included statistics and, by drawing on the five articles, write an opinion piece on the proposed new law.

The activity engages students with a relevant text type (newspaper) to create a persuasive text, thereby addressing a goal of the English curriculum. From a numeracy perspective, the proposed new dog laws provide a meaningful context for students to employ mathematical knowledge (find and interpret statistical information) that is presented using representational tools (tables and graphs). The teacher's goals include promoting students' dispositions towards using mathematics by choosing a subject likely to be of interest to them. Furthermore, students are encouraged to adopt a critical orientation by making decisions about the sources of the data included in the articles to justify their position.

> **Review and reflect 5.2**
>
> For this task, work with a partner who has a specialisation or interest in a different learning area from you.
>
> 1 Read the following three short articles, which provide further examples of activities that promote both numeracy and subject learning: Cooper et al. 2012; Gibbs et al. 2012; Willis et al. 2012.
> 2 With your partner, choose a task from each of the articles. Identify relevant curriculum goals in the Australian Curriculum and numeracy demands that have been highlighted by the numeracy icon. What additional numeracy opportunities did the teachers find?
> 3 Discuss your findings with another pair of colleagues.

History

The Australian Curriculum: History includes two interrelated strands: 'Historical knowledge and understanding' and 'Historical skills'. Students in each year level investigate a particular historical context guided by a set of key inquiry questions. Study of developments within the historical context provides an opportunity to develop historical understanding and skills.

One of the historical developments that students in Year 8 can investigate is the Spanish conquest of the Americas (c.1492–c.1572). The numeracy icon does not appear beside any content elaborations for the investigation, suggesting that there are no inherent numeracy demands associated with this content (see Figure 5.3).

To show how teachers can create relevant numeracy learning opportunities in history, we will draw on three case studies that come from lessons taught by the same teacher on the short-term and long-term effects of the Spanish conquest of the Americas. The first case study utilises tasks that can help students develop historical understanding, while the focus in the second and third case studies is on a historical skill. Together, these case studies illustrate how small changes in the way an activity is implemented can enrich a task so that it creates numeracy opportunities beyond the inherent demands of the published curriculum.

Historical concepts and numeracy

The key concepts for developing historical understanding include 'evidence, continuity and change, cause and effect, perspectives, empathy, significance and contestability' (ACARA, 2018d). Many of these concepts are abstract and can be challenging for students to understand. Phillips (2002) argued that explicit attention to numeracy has potential to make abstract historical

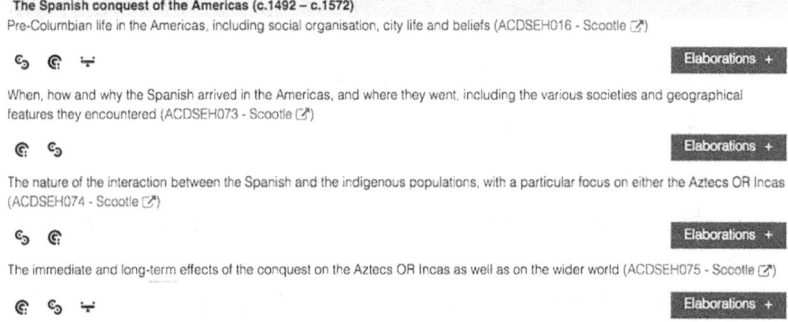

Figure 5.3 Excerpt from the 'Spanish conquest of the Americas' depth study of the Australian Curriculum: History for Year 8

Source: ACARA 2018d.

concepts concrete. He illustrated this claim by utilising a series of calculations designed to help students understand the historical significance of the Atlantic slave trade and why it endured for so long despite moral arguments against slavery. The data he provided included the expenses and income for a single voyage of a ship from a Liverpool port. When combined with information about the number of voyages from Liverpool ports and the amount of money injected into the Liverpool economy from crews' wages, the profitability of the slave trade is clear. This information helps students understand the economic impact of the slave trade (significance) and its longevity (cause and effect). From a numeracy perspective, students employ numerical calculations (mathematical knowledge) to understand aspects of the Atlantic slave trade (context). The data are presented in a table (representational tool), and students have the opportunity to adopt a critical orientation by asking questions of the data (e.g., Were there hidden costs? How accurate are the records?). Case study 5.3 further illustrates how identifying and explicitly attending to numeracy opportunities have potential to increase students' understanding of abstract historical concepts.

Case study 5.3 Understanding the Spanish conquest of the Americas

Students were shown a map of South America in which a shaded region indicated the location of the Incan empire (see Figure 5.4). The teacher asked students why the Incan empire was located in this region—that is, along a narrow strip of land on the western coast of South America.

During the class discussion that followed, some students suggested that the sea, rainforest and mountains that

Figure 5.4 Map of South America showing the location of the Incan empire

bordered the shaded region determined the location of the Inca. These students had drawn on their prior knowledge of the geography of South America, as the rainforest and mountains are not shown on the map. A topographical map would have enabled students without this prior knowledge to gain an appreciation of the height of the mountains and the locations of the rivers, thereby helping them to understand why the Incan empire was located in this particular region. The teacher noted this additional numeracy opportunity, which she had been unable to take advantage of due to time constraints, when discussing this task in a post-lesson interview: 'I would have liked to look at a map in relation to other people groups. So how big were the Incas in relation to the Aztecs and the Mayans... Looking at the topographic maps, where were the rivers? Where were the lakes? Where was the coastline? Where were the mountains? That would be really interesting'.

The topographical features of the landscape provide a reason for the location of the Incan population, thereby addressing the historical concept of cause and effect. Mathematical knowledge (scale and measurement) in conjunction with the capacity to interpret a representational tool (the map) is necessary to understand why the geographical features of the landscape created boundaries for the Incan empire.

In the following lesson, the teacher provided students with population data for Inca, Spanish and African peoples in South America from 1491 to 1600 (see Table 5.2). The teacher then orchestrated a class discussion directed by the following prompts:

- Pre-Columbian population numbers are almost impossible to estimate accurately. Suggest reasons why.
- Examine population estimates in the table. Describe the patterns over time for each group. (Does the population increase or decrease, and by how much?)
- Suggest reasons for patterns of change for the different populations.

Questions

You may need to visit a website that provides information about the Spanish conquest of the Americas before you undertake this task. With a partner, discuss your responses to the three prompts and the key historical concepts that

Table 5.2 Population data for the Inca, Spanish and African peoples in South America, 1491–1600

Year	Inca	Spanish	African
1491	50–70 million	0	0
1550	30 million	250,000	50,000
1600	10 million	700,000	250,000

could be addressed. Analyse the task in terms of the dimensions of the 21st Century Numeracy Model.

The class discussion began by considering the scale of the reduction in the population of the Inca in the period from 1491 to 1600; that is, by 1600 the Inca population was approximately one-sixth of the population estimate prior to the arrival of the Spanish. To give students some sense of what this meant, the teacher related the reduction in population to the number of students in the class: 1 in 6 Inca survived, so only 4 of the class of 24 students would survive.

The discussion then turned to why it was impossible to accurately estimate the population of the Inca before the arrival of Columbus. Suggestions from students included that no accurate census had been conducted before that time, records had been destroyed, and loss of language meant that any surviving records could not be interpreted. The teacher did not pursue any of these suggestions; however, they provided additional numeracy opportunities. For example, there was an opportunity to discuss what a census is. This is an example of a *numeracy moment*: an unplanned opportunity to address an aspect of numeracy.

Turning to possible reasons for the changes in population sizes of Inca, Spanish and African peoples in South

America, the teacher reminded students that in Central America 80 per cent of the Aztecs had died from diseases brought by the Spanish. Students were able to speculate on possible reasons for the increase in Spanish and African populations: the discovery of resources and the need for labour for the fields and mines, respectively.

The teacher's goal was for this task to help students to understand the impact of the Spanish invasion on the Incan empire and to begin to make connections between what was happening in South America and the slave trade:

> So we went through and the table just showed that it went from 50 to 70 million people of the Indigenous population, so the Incas went down to 10 [million] a century later and then the other [columns] looked at the rise in population of the Spanish and the African slaves because we haven't really, we've been touching on the African slave issue ... I was trying to lead them into that and I'm sure we'll cover it next week in class. Just the concept that there was a reason why they were there.

The lesson addresses the historical concept of cause and effect by drawing on evidence in the form of population data. Additional questions can be asked: Why do people in South America speak Spanish? Why are Black Americans located predominantly in the southern states of the United States? From a numeracy perspective, the data are presented in a table (representational tool) from which students are required to use their mathematical knowledge to extract relevant information. By asking students to consider why it was impossible to estimate the number of Inca before the arrival of the Spanish, the teacher was asking students to adopt a critical orientation: what is the source of the data? This question also addresses the historical concept of contestability, which results from different interpretations of the same event and can arise from using different sources of evidence or taking different perspectives. Thus, there is close alignment between the historical concept of contestability and the numeracy dimension of adopting a critical orientation.

Historical skills and numeracy

There are five historical skills to be addressed in Years 7–10: 'Chronology, terms and concepts', 'Historical questions and research', 'Analysis and use of sources', 'Perspectives and interpretations' and 'Explanation and communication'. Only the first of these is identified by the numeracy icon as having inherent numeracy demands: 'Use chronological sequencing to demonstrate the relationship between events and developments in different periods and places'. However, the examples in Case study 5.3 illustrate how numeracy can help students develop the other historical skills.

Understanding in history involves thinking about the past as having structure and direction; therefore, an understanding is needed of chronological conventions and 'temporal concepts like

Review and reflect 5.3

Revisit Case study 5.3 and discuss how the lessons help students develop the historical skills listed in the Australian Curriculum: History.

"now", "then", "before", "after", "sequential", "concurrent in time (= instantaneous)" and "over time (= has duration)"' (Blow et al. 2012, p. 26). These concepts involve sequencing and scale and thus involve numeracy. A timeline is a representational tool referred to in every year level of the Australian Curriculum: History—but only to sequence historical events. A comparison of the following two case studies highlights other aspects of time that are important for developing historical understanding.

Case study 5.4 Features of timelines

During a class discussion, students were able to provide many of the features of a timeline (e.g., title, scale, dates, equal spaces between decades, an abbreviated description of the events). Their subsequent task was to place five historical events that took place during the Spanish conquest of the Americas in chronological order on a timeline (see Table 5.3). A volunteer drew his timeline on the whiteboard but did not include all the features that had been identified previously.

Table 5.3 Selected events during the Spanish conquest of the Americas

Events	Years
Cortés captures Tenochtitlan	1519
Columbus discovers the New World	1492
Magellan's ships circumnavigate the globe	1520
Balboa reaches the Pacific Ocean	1512
Pizarro invades the Incan empire in Peru	1531

Case study 5.5 Constructing timelines

The teacher orchestrated a whole-class discussion about the important features of a timeline through questioning. The questions she asked included: Why do you need to measure? How far apart in time were the events? Why do we use an arrow? Why can't we just have 'Timeline' as the title? She summarised the important features of timelines on a PowerPoint slide, emphasising the need to use a ruler and an appropriate scale. Students were asked to construct a timeline for some events that occurred during the Spanish conquest of the Americas (see Table 5.4).

Table 5.4 Significant events and developments during the Spanish conquest of the Americas

Events	Years
Christopher Columbus sails west from Spain and arrives on the Caribbean Islands	1492
Bartolomeu Dias sails south along the west coast of Africa to the Cape of Good Hope	1488
Ferdinand Magellan sails west from Spain and completes the first circumnavigation of the globe	1522
Prince Henry of Portugal establishes the first European school of navigation	1418
Vasco Núñez de Balboa travels west overland in the Americas to Panama	1513
Vasco da Gama sails south and around the Cape of Good Hope, then east, becoming the first European to reach India by sea	1498

Students were allocated to groups after they had constructed their timelines and given a page of text about

> Columbus, Balboa, Cortés or Pizarro. Their task was to locate the explorer's country of origin and date of birth, where he went, his most significant achievement and the year and cause of his death.

Case studies 5.4 and 5.5 illustrate extremes on a continuum that ranges from presenting a timeline and asking students to copy it into their notes to choosing information that highlights sequencing, concurrency and duration of events. At one extreme, a teacher would be addressing the numeracy demand of using a timeline to sequence events, while at the other they would be enriching historical understanding by incorporating scale so that concurrency and duration can be represented.

In Case study 5.4, the teacher briefly touched on the features of a timeline, but she addressed them more explicitly in Case study 5.5. However, the historical events in both case studies were listed as occurring in single years, with no two years being the same, and so the appearance of these events on a timeline is as discrete points. Historical events occur over periods ranging from a particular date to years, and some events are concurrent. By choosing what is to be included on a timeline, teachers can move students towards understanding the concept of time and abstract historical concepts such as continuity and change, and cause and effect. For example, the information provided to students in Case study 5.5 after they had constructed their timeline could have been used as the basis for selecting and representing events on the timeline. Furthermore, the construction of a timeline in this way illustrates how contestability

> **Review and reflect 5.4**
>
> Timelines are representational tools that are used in several different learning areas.
>
> 1 With a partner who teaches the same subject as you, choose a year level and identify where a timeline might help develop understanding of a disciplinary concept.
> 2 Analyse the use of timelines in terms of the dimensions of the 21st Century Numeracy Model.
> 3 Share your findings with colleagues who are preparing to teach different subjects.

arises in history, because any timeline only shows selected events for a particular period.

Analysing tasks for curriculum goals and numeracy

The case studies presented in this section illustrate how teachers were able to identify numeracy opportunities and then design activities to simultaneously address curriculum goals and promote numeracy learning. We have explicitly linked the activities to curriculum goals in various learning areas in the Australian Curriculum and analysed them through a numeracy lens. To plan for numeracy opportunities, as these teachers did, it is necessary to view potential tasks through both a disciplinary and a numeracy lens, by asking two questions: How does the task contribute to disciplinary understanding? and What dimensions of numeracy does the task help students develop? Figure 5.5 is a template we developed to help teachers to analyse tasks retrospectively, but it can also be used to plan for numeracy opportunities.

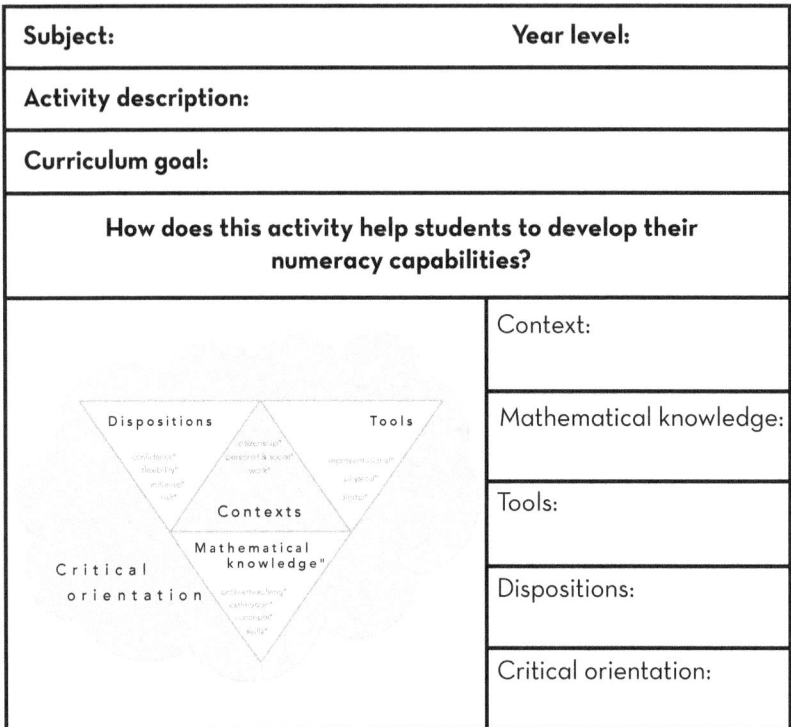

Figure 5.5 Template for analysing numeracy tasks

Note: Not all dimensions of numeracy will be addressed in every task but are likely to be addressed over the course of a unit of work.

NUMERACY OPPORTUNITIES

How can you plan to take advantage of numeracy opportunities in the subjects you teach? Do you start with a curriculum goal, or do you start with a numeracy-focused task? The answer to the second question is that either starting point will work. However, when you are beginning to develop the capacity to pay explicit attention to numeracy, it may be easier to get started by thinking about the four aspects of numeracy that Crowe (2010, p. 106) identified as being fundamental for informed citizenship, albeit in the context of the

> **Review and reflect 5.5**
>
> 1. Visit the University of Queensland (2018) page on the Filmpond website.
> 2. Four of the videos on this page come from lessons across different year levels and subjects: *Year 1 Mathematics*, *Year 9 German*, *Year 9 Design and Technology* and *Year 10 Health*. View one of the videos and explore the published curriculum for that year level and subject.
> 3. With a partner, discuss how the task in the video relates to curriculum goals and analyse the task using the 21st Century Numeracy Model. Record your findings in Figure 5.5.
> 4. Share your ideas with another pair of colleagues who viewed a different video.

social studies curriculum: '(1) the ability to understand *raw numeric data* in context, (2) the ability to *understand percentages* in context, (3) the ability to understand the *meaning of average* in context, and (4) the ability to *interpret and question graphs and charts*' (original emphasis).

CONCLUSION

In this chapter, we explored how teachers can identify numeracy opportunities that enhance disciplinary learning. We also introduced the notion of a numeracy moment—an unplanned numeracy opportunity that arises in the normal course of a lesson. We modelled how tasks can be analysed to identify curriculum goals

> **Review and reflect 5.6**
>
> 1 Read Crowe's (2010) article on numeracy in social studies.
> 2 With a partner who teaches the same subject as you, choose a year level and content description from the Australian Curriculum where one of the aspects of numeracy identified by Crowe is relevant.
> 3 Find a resource for this content description (e.g., from Scootle, <www.scootle.edu.au>) that addresses this aspect of numeracy. Use Figure 5.5 to record your analysis of the task.
> 4 What other aspects of numeracy exist in the subject you teach? Compile a list of topics and relevant aspects of numeracy. (Consider the four aspects of numeracy identified by Crowe and any others you can see.)
> 5 Compare your list with another pair of colleagues who teach a different subject. Identify similarities and differences.

and dimensions of the 21st Century Numeracy Model, thereby making explicit connections between subject and numeracy learning. Finally, we stressed that many aspects of numeracy are evident in more than one learning area. Consequently, these aspects of numeracy provide a useful starting point for developing the capacity to see numeracy in the subjects you teach.

RECOMMENDED READING

Cooper, C., Dole, S., Geiger, V. & Goos, M., 2012, 'Numeracy in society and environment', *Australian Mathematics Teacher*, vol. 68, no. 1, pp. 16–20

Crowe, A., 2010, '"What's math got to do with it?" Numeracy and social studies education', *The Social Studies*, vol. 101, no. 3, pp. 105–10, doi: 10.1080/00377990903493846

Gibbs, M., Goos, M., Geiger, V. & Dole, S., 2012, 'Numeracy in secondary school mathematics', *Australian Mathematics Teacher*, vol. 68, no. 1, pp. 29–35

Peters, C., Geiger, V., Goos, M. & Dole, S., 2012, 'Numeracy in health and physical education', *Australian Mathematics Teacher*, vol. 68, no. 1, pp. 21–7

Phillips, I., 2002, 'History and mathematics or history with mathematics: Does it add up?', *Teaching History*, vol. 107, pp. 35–40

Willis, K., Geiger, V., Goos, M. & Dole, S., 2012, 'Numeracy for what's in the news and building an expressway', *Australian Mathematics Teacher*, vol. 68, no. 1, pp. 9–15

6

Planning for numeracy across the curriculum

To develop students' numeracy capabilities effectively, teaching and learning experiences must be carefully planned. This planning includes the design of tasks that incorporate the dimensions of numeracy—contexts, mathematical knowledge, dispositions, tools and critical orientation—and the pedagogies employed to optimise student learning. In Chapters 4 and 5, we established the difference between numeracy demands and opportunities and developed ways in which these could be identified in curriculum documents (e.g., Goos, Dole et al. 2012). This chapter will explore approaches to planning single lessons and lesson sequences based on thoughtfully designed tasks aimed at promoting numeracy capability as well as competence within relevant curriculum learning areas. We will also discuss how these tasks are brought to life in classrooms through appropriate pedagogical approaches.

The chapter begins with a discussion of the general principles of task design. Then, we outline how effective teachers of numeracy think of ideas that form the basis for numeracy tasks and structure such tasks. This is followed by a description of appropriate strategies for the implementation of tasks in the classroom. Finally, we provide examples of rich numeracy tasks used by teachers we have worked with.

PRINCIPLES OF TASK DESIGN

While it might seem that good teaching ideas just 'pop into your head', there has been considerable research into how to design effective tasks for student learning. Burkhart and Swan (2013) have argued that tasks are integral to many dimensions of mathematics learning, including content, processes and modes of working. There is also evidence that carefully designed tasks are effective in improving teaching practice, through the success of long-term programs such as Connected Mathematics (Lappan & Phillips 2009).

Teachers may select tasks from available resources and often adapt such tasks to match the specific learning needs of their students. Sometimes, teachers also create tasks for themselves. In each of these cases, underlying criteria or principles of task design are used as the basis for selection, adaptation or creation, so that the best task for a particular teaching scenario is used. Further, the task must be connected to the choice of pedagogy employed in order to gain the full benefit that lies within it (Sullivan & Yang 2013).

As most tasks are developed for implementation within specific curriculum and school contexts, the *fit to circumstance* of tasks with local conditions and constraints is important for effective implementation (Kieran et al. 2013). Such circumstances include local curriculum characteristics as well as other considerations such as the teaching resources available within a school. The fit to circumstance of tasks also includes the choice of pedagogy that best supports the learning of the particular group of students at which it is aimed. Diezmann et al. (2001, p. 170) define mathematical investigations as 'contextualized problem solving tasks through which students can speculate, test ideas and argue with others to defend their solutions'. As this definition aligns with the development of numeracy capability, it suggests that the adoption of investigative pedagogies is important in order to take full advantage of the numeracy opportunities and demands embedded within a task (Goos, Geiger & Dole 2013).

Challenge is important for students if real learning is to take place (Hiebert & Grouws 2007). Most guidelines for improving learning outcomes stress the need for teachers to extend students' thinking by posing extended, realistic and open-ended problems (e.g., City et al. 2009). By posing such challenging tasks, teachers provide opportunity for students to take risks, to justify their thinking and to work with other students (Sullivan 2011). Challenge also includes opportunities for students to make decisions and judgements and so exercise and develop their capacities to use mathematics critically (Geiger, Goos & Dole 2014).

Students, however, often resist engaging with challenging tasks and attempt to influence teachers to reduce the difficulty of an activity (Sullivan et al. 2013). Thus, while it is important for students to engage with learning experiences that offer challenge, tasks must also appear to be achievable—that is, challenging yet accessible. For tasks to be accessible they must be transparent; that is, it must be clear what students are expected to do, and there must be points of entry where every student can begin an activity (Burkhart & Swan 2013).

For the quality of a task to be assured, the task's activities must also be trialled, evaluated and re-trialled in cycles of design and improvement (Maass et al. 2013). Thus, effective activities will take time to develop and require a commitment to reflection and improvement by teachers as designers of tasks.

PRINCIPLES FOR NUMERACY TASK DESIGN AND IMPLEMENTATION

This section outlines a set of principles for the design and implementation of numeracy tasks. These principles were developed by one of the authors of this book while working with ten teachers within a project using the general principles of task design as a starting point. The teachers had both primary and secondary teaching backgrounds, and half had been involved in previous numeracy projects. The framework is presented in three sections: *identifying* an idea that could be used to meet a numeracy demand or opportunity, *shaping* the idea into a task, and *actualising* the task in a classroom through an appropriate pedagogical approach.

Identifying an idea

Thinking of new ideas for numeracy is not a trivial exercise, as it requires creativity, a deep knowledge of curriculum objectives and the ability to bring these two aspects of teaching together. The first step in identifying an idea that has potential for a numeracy task is to develop a disposition to be always *looking* for sources that align with the numeracy demands or opportunities of a teaching and learning sequence. Thus, looking is a sensitivity or openness to ideas that could be brought into the classroom in the form of numeracy tasks. Once looking, effective designers of numeracy tasks begin *noticing* real possibilities for potential numeracy tasks—an event, phenomenon or experience that takes place inside or outside school that might form the basis for a task. Noticing can lead to selecting or adapting existing school activities or resources, or creating new activities. To bring good ideas into the classroom, however, effective teachers of numeracy must also start *seeing* how an initial idea for a task aligns with curriculum documents or school-based teaching and learning programs. This process allows teachers to plan when an activity will fit into a program and also decide how the initial idea will need to be adapted or shaped to match curriculum objectives for both numeracy and the learning area in which an activity will be introduced.

Case study 6.1 Looking, noticing and seeing

Through the curriculum

Olive was an early childhood teacher working in a school within the Catholic education sector. The school was located in a satellite city 45 kilometres from a state capital. Olive had developed a series of tasks for her group of preparatory students (typically five to six years of age) as part of an activity rotation that integrated mathematics and religious studies (see Geiger 2016).

Olive's school had recently acquired an adjacent block of land on which they were planning to build additional classrooms as part of a school expansion. While plans for new buildings were underway, Olive had asked permission to involve her students in creating a prayer garden in the back yard of the recently acquired property. As part of this project, she developed a series of tasks that matched the curriculum objectives of a number of learning areas.

In one of these tasks, students were asked to determine if a long rectangular bench seat could be moved to a different position within the garden. They were not permitted to discover the answer by moving the seat itself, as it was too heavy for preparatory students. No formal measuring tools (tape measures, rulers) were available, as the students had not yet learned about formal units of measure. Instead, square tiles measuring 30 · 30 centimetres were provided as informal units of measure.

After the teacher had explained the task, students discussed among themselves how they could go about the activity. After this discussion, students used the tiles to determine a measure of the bench seat's length by placing the tiles end-to-end along the top of the seat. Once the length of the bench seat had been covered, the tiles were gathered up and moved by the students to the proposed new site for the seat. Starting at one end of the designated space, students laid out the tiles end-to-end on the ground until they reached the other end of the area in question. There were leftover tiles, which were then piled up at the end of the space. The students concluded that there was not enough space to move the bench seat to the proposed space and that another place would need to be found.

After the lesson, Olive was asked how she had devised the task. She said that she had started by reflecting on the content of relevant curriculum documents she needed to address at that time of the year. At the same time, she had also considered what resources or aspects of the environment could be utilised. Eventually, she had brought curriculum and the resources offered by the built environment together. Olive believed it was her familiarity with curriculum documents that allowed her to pick out relevant strands from both mathematics and religious education and bring these together through the opportunity made available by the purchase of the new property. When asked if this was typical of the way she developed numeracy tasks, she replied that she had worked hard to be thoroughly familiar with the curriculum requirements for any year level she was teaching, and she used this as a lens when looking to create new activities.

By archiving ideas

Richard was an English teacher in a government secondary school situated in a regional centre. In a Year 9 lesson that he conducted, students were required to write a letter to a new pen pal who lived on Horn Island, located off the northern Australian coastline in the Torres Strait. As preparation for this assignment, Richard had asked his students to research a number of aspects related to the island, including its population, land area and number of schools, the distance from their home to Horn Island, available means of transport and the travelling time from their home to Horn Island and the frequency of transport to and from the island. Richard had asked students to gather this information in order to help them understand the life circumstances of their new pen pals and so provide them with starting points for their first letter.

Students worked enthusiastically on this task through the lesson, regularly expressing surprise at what they found; for example, the population of Horn Island was only 539 people, a small fraction of the population of the country town in which the students lived. The lesson concluded before the task was completed, with Richard telling students it would be continued the next day.

When asked how he had come up with the idea for the lesson, Richard replied that he was always on the lookout for opportunities to promote students' understanding of the challenges faced by others in the world. This was part of his outlook on life, as he had a strong sense of social justice. Richard described how he had seen a television documentary about Horn Island and had 'parked' it as the source of an idea he could use in his classroom at some future date. He remembered the idea when looking forward through his teaching program for the semester and made the connection with the idea he had archived earlier and the English unit on writing to a pen pal. This provided opportunity for him to bring into the classroom an aspect of teaching he valued—the promotion of tolerance and social justice through an understanding of the circumstances of others—and a curriculum requirement.

In this case, Richard had not used the curriculum as a lens through which to look out into the world for a teaching idea but rather as a framework, or overarching plan, to which he could attach teaching ideas he had already identified as having potential at some future time. In Richard's case, the potential he saw in the identified idea was based on his belief in the importance of promoting students' awareness of the circumstances of others.

> **Questions**
> 1 Describe the ways in which Olive's and Richard's approaches to finding new numeracy teaching ideas are the same or different.
>   - What role did a knowledge of curriculum play in either case?
>   - How were the ways in which they planned different?
> 2 Discuss with a partner your current approach to developing numeracy teaching ideas.

Shaping a task

After an initial idea for a numeracy task is identified, two processes are involved in fully developing the task: *structuring* and *fit to circumstance*. To structure tasks, the 21st Century Numeracy Model can be used to check that the chosen context would engage students, relevant mathematical knowledge is identified, consideration is given to how the activity would promote positive student dispositions, appropriate tools are introduced, and elements are included in the task to challenge students to adopt a critical orientation. In creating, selecting or adapting tasks, teachers also seek to accommodate or take advantage of their school's unique characteristics—to fit to circumstance. In doing so, teachers consider:

- the specific learning needs of their students (e.g., the number of students who are non-dominant English-language speakers)
- their school's specific focus within curriculum or local interpretation of learning area documents (e.g., literacy, numeracy, self-confidence, development as an independent learner)

- available teaching and learning resources within the school (e.g., commercial learning aids, books of teaching and learning activities)
- the potential of their school's built environment as a resource (e.g., swimming pool, school ovals)
- the potential of the local natural environment as a resource (e.g., nearby waterways, proximity to historical landmarks)
- how to manage the introduction of new activities when working with colleagues (e.g., convincing colleagues to implement a task for all students in a particular year level).

Case study 6.2 Structuring and fit to circumstance

Kym was a primary teacher within a school in a regional town. In developing a unit in mathematics for her Year 6 students, she took advantage of a potentially disruptive new building construction in her school—a consequence of a major government initiative supporting school development (see Geiger, Goos & Dole 2013). A description of the task as it appeared in Kym's planning documents is presented below:

> The construction at the school has caused many changes. The school map is now out of date and new parents would get confused trying to use it. Your task is to alter the out-of-date map so that it reflects the changes to the school. Then design a tour of the school for new parents that will show them where everything is.
>
> You need to make your tour on the map. Write a tour guide to go along with it so that parents can do

the tour alone, with instructions like, 'On your left you will see the parent entrance to the office. If you look to your right you will see the Years 6 and 7 building.'

The tour is to be no longer than 10 minutes. When you have designed your tour, test it out on a group to check for errors and timing.

Before final drafts, compare journeys. Discuss differences in routes, school highlights and times taken for the journey. Discuss any glitches that came up in organising tours. Precision in map reading and instructions is important so people don't get lost.

To begin the task, students were given a plan of the school before the construction began and asked to label key landmarks. They were then asked to redesign the plan to include the new buildings within the school. This was to be a scaled plan that included a path around the school without getting too close to the buildings under construction. The plan was to be marked with distances and locations.

The task also required students to develop instructions for a tour of the school as part of an induction for preparatory students' (four to five years of age) parents. These instructions needed to accommodate the new buildings. Students' plans and instructions were then swapped with those of other students to test their effectiveness. The experiences of using other students' directions were shared in a whole-class discussion.

Kym reported a high level of engagement of students with the task. She argued that the investigative approach she adopted allowed students an element of control over their own learning, which contributed to this engagement. Kym also believed that there were two more contributing factors to students' engagement: the outdoor nature of the activity and the relevance of the task to students' personal circumstances. 'Students have control of where they want to take it. The students tell me this approach allows them to look into "stuff that we want to learn about".'

In structuring the task, Kym had chosen an engaging context based on building redevelopment within the school. Mathematical knowledge was addressed, as students were required to construct scale maps and make accurate use of the language of location in order to provide directions for other students in their role as school tour guides. Kym attempted to promote students' positive dispositions towards the use of mathematics by introducing a context that students found relevant and that challenged them to think flexibly and adaptively. Students were also required to make use of representational tools such as maps, and Kym made use of digital tools herself through the use of Google Earth to provide students with a plan view of the school. The whole activity was embedded in a critical orientation, as students had to make decisions and judgements related to the best route for a school tour.

Questions

1. Kym's task clearly aligns with the structuring aspect of the framework, but what did she consider in relation to fit to circumstance? Make a copy of the list of fit to circumstance aspects at the beginning of this section and describe how each aspect is apparent in Kym's task.
2. Do you think it is possible to include all of these aspects in any numeracy task?
3. Are some aspects more important than others?
4. Discuss your position with a partner.

Actualising a task

Specific teacher capabilities that are necessary for the delivery of effective lessons embedding numeracy across the curriculum represent the *pedagogical architecture* of the lessons. These capabilities include:

- initial setup of the lesson, which involves explaining the task and building students' understanding of the context in which the activity is set. This phase usually includes a *critical question* students are to investigate through the lesson. The purpose of this part of the lesson is to make the task both transparent and accessible to students as well as providing challenge and direction for students' critical orientation.
- initial selection of pedagogy (or pedagogies) appropriate for the task. These usually include a degree of teacher-directed activity, but the intention is usually to adopt an investigative approach for the main part of the lesson.
- making flexible use of a repertoire of pedagogical strategies in order to adapt to unforeseen events during a lesson. Teachers call this the ability to *flip*.
- ability to adapt tasks on the fly in case of unanticipated student responses.
- use of a *measured responsiveness* to student questions—i.e., providing just enough information or feedback for students to continue with a task without removing challenging aspects of an activity.
- conclusion of lessons where students are brought together to review what they have learned. This phase has a focus on the

original critical question. In responding to this question, teachers insist students provide evidence as part of the justification of their views.

> ### Case study 6.3 Pedagogical architecture
>
> Tanya was a secondary mathematics teacher in a large city. In this lesson, she worked with a Year 7 mathematics class on a problem set in a real-world context using an investigative approach. In addition to providing students with experience in mathematical problem-solving, her intention was for students to practise the calculation of and conversions between fractions and ratios in order to meet the objectives of the school mathematics program at that time.
>
> The lesson began with Tanya posing the following question to the class: What is a typical Year 7 student? Tanya then explained that by *typical* she did not mean average but instead was referring to common ratios between body parts—for example, shin length compared to leg length. In other words, students were asked to establish typical body part ratios for members of the class. Tanya asked that ratios be expressed as fractions reduced to their lowest terms. In order to make comparisons easier, she asked students to consider recording their fractions in simpler terms as approximations; for example, 22/85 could be approximated as ¼.
>
> Students were required to calculate the ratios, fractions and simpler approximate fractions for the following body parts:
>
> - foot and shin length
> - foot and leg length

- index finger and hand length
- hand and forearm length
- hand and foot length
- arm and leg length
- head and torso circumference
- torso circumference and leg length
- foot and forearm length
- height and leg length
- height and arm length.

Tanya asked students to work in groups of three to four and provided them with rulers and measuring tapes. She spent time ensuring students understood the question and explained that a table had been drawn on the whiteboard for recording their foot to leg length ratio as a start to the activity.

Students appeared happy to engage with the activity, offering 5/8, 22/93 and 22/85 as answers to the initial body part ratios. Tanya then asked students if they could convert 22/93 into a simpler approximate fraction so it could be compared with 5/8 and 22/85. It soon became apparent they were struggling with this idea, so after several attempts to explain how to find relevant approximations, she abandoned this approach, realising it was a distraction from the overall aim of the lesson and that the conversion idea could be dealt with at a later time.

Instead, Tanya decided students should convert each proper fraction into a decimal fraction in order to compare body part ratios. They began by writing 0.625, 0.24 and 0.26 on the whiteboard. These initial calculations made Tanya wonder how the first clearly unusual ratio had been determined. After questioning the class, she found other students had recorded similar results. It was at this point she also noticed a pair of students, a little behind the others, working together to measure the length of one of their legs—while in a squatting position. Tanya stopped the class, asked for everyone's attention and began a whole-class discussion about why there were such variations in results. In conducting this conversation, she was careful not to just tell the students where the problem lay, instead offering prompts such as 'How do you measure foot length?' and 'Show the class how you measured the length of your leg'. This type of prompting continued until one of the students pointed out that some of her class members were measuring foot and leg length in different ways. This comment began a whole-class discussion about the correct way to measure different body parts and the need to apply consistent methods for an activity of this type. It had not previously occurred to Tanya that the correct method for measuring body parts would be an issue, as she had assumed all students would have learned to do so previously. After checking all students were now aware that leg length was measured while standing up and that they also knew how to measure the other listed body parts, Tanya asked them to return to the investigation and complete the required calculations.

At the end of the lesson, Tanya questioned her students about what they had achieved over the previous 45 minutes, bringing them back to the question she had proposed at the beginning of the session. During this discussion, she worked with her students to establish what they needed to do in the next lesson to provide an adequate response to the critical question.

Questions

Work with a partner to match aspects of the lesson to the pedagogical architecture described by addressing the following questions:

1 How did Tanya organise the initial setup of the lesson?
2 Identify the critical question.
3 What was Tanya's intended pedagogy at the beginning of the lesson? How did this change during the lesson as she drew on other aspects of her pedagogical repertoire of teaching approaches?
4 Why and how did Tanya change the task during the lesson?
5 How did Tanya respond to her students when their results were unusual?
6 How was the lesson brought to a conclusion?

Case study 6.4 A numeracy task to support learning in English

This case study is concerned with a Year 8 (14-15 years of age) English lesson aimed at developing students' appreciation for the use of pace when reading poetry. The particular focus of the lesson was on improving students' oral presentation skills by promoting an understanding of the relationship between the emotions being communicated in a poem and the associated pace at which different sections of a poem should be read.

Kathy began the lesson by explaining to students that *pace* in this instance is the speed at which poetry is read. She went on to explain that pace is varied according to the context of the ideas or events explored in a poem and is closely tied to the emotions a poet is attempting to invoke in a reader. This began a discussion among the class about how to vary pace and how this could be measured. Students made a number of suggestions, including beats per minute, number of words per unit of time (e.g., minutes, seconds) and syllables per unit of time. To follow up on these suggestions, Kathy invited one of the students to read a short poem by Rick Roth (2011) 'So fast'.

> So fast, so fast
> an eye's quick blink
> had always heard
> but didn't think
>
> it possible for all to go
> so fast
> more fast
> than I could know
>
> I wish I'd stopped
> to linger more
> to take it in
> and feel the core
>
> of that which mattered
> most to me
> I didn't see
> I didn't see
>
> that folks would age
> and babes would grow
> friends would travel
> to and fro

The times I loved
fleeting they were
too many now
become a blur

and as I contemplate the past
I wish it hadn't
gone so fast
and that today would better last

While the poem was being read, a group of students counted the number of syllables in the poem, and another group timed how long it took for the poem to be read. At the end of the reading, students reported back that there had been 120 syllables read in 24.4 seconds—approximately 5 syllables per second. After a class discussion, students concluded that the pace at which the poem had been read aloud was too fast for the emotions they felt should be conveyed through the poem. This conclusion prompted Kathy to ask what emotions students felt should be communicated within the poem. Students suggested a number of emotions, including regret, sadness and anger.

Kathy then asked the students how emotions might be matched to pace, leading to further discussion about how fast the poem should be read. Students eventually reached a consensus that the reading should reflect 'slow for sad and faster for angry'. At this point, Kathy asked three students to come to the front of the room, with the first to read a section of the poem in a sad way, another to read at a pace related to feelings of anger and the third reading in an excited manner. When they had finished, Kathy asked the three students to stand across the front of the room in an order related to the pace at which they had read the poem. The student who delivered a 'sad' reading of the poem stood to one side of the room, while the 'angry' and 'excited' readers stood to the other side.

The class was then asked by Kathy to name emotions that should be related to a medium pace, to which two students replied 'happiness' and 'boredom'. These students were also invited to stand at the front of the room at an appropriate spot between the two students representing fast and slow; however, the remainder of the class were not convinced these represented 'middle' emotional positions. After further discussion students eventually settled on the word *fine*, and the student who made the suggestion was asked to stand in the middle of the line of students at the front of the room. Members of the class then made suggestions about the relative positions of the students in relation to each other. This led to some rearrangement before students settled into a final position. The students at the front of the room were then asked to write the emotion they represented on a sticky note and place this on the whiteboard by reaching directly behind them. Kathy drew a line parallel to the sticky notes and annotated the line by placing scale marks against each sticky note and writing the emotion in larger print so the whole class could inspect the position of each named feeling. After again checking with the class that these emotions were appropriately placed, Kathy asked what emotions were missing. Students provided suggestions for additional feelings, such as lazy, maudlin, relaxed and sickened, which were also marked on the whiteboard (Figure 6.1). Kathy explained to the class that they had developed a way of quantifying emotions by placing them on a line and added the title 'Emotion scale' to the diagram.

After assisting students to develop the emotion scale, Kathy introduced a table for them to use when describing

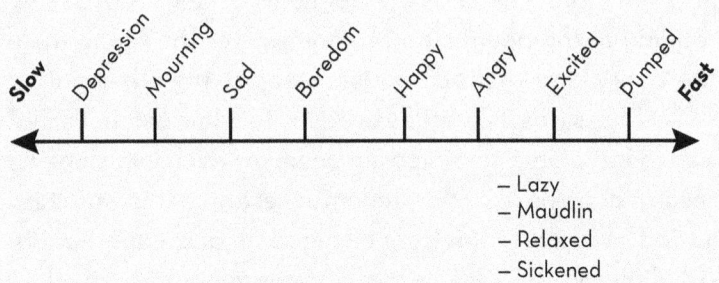

Figure 6.1 Scale relating emotions to the pace of reading a poem

the pace at which a poem should be read. The table was structured around the headings of 'Poem', 'Emotion', 'Evidence' and 'Pace'. Students were asked to complete the table for 'So fast'. After the students made an attempt to do so as individuals, Kathy orchestrated a discussion with the class that resulted in a completed list displayed on the whiteboard in which the pace chosen for the poem was moderately slow:

Poem   'So fast' by Rick Roth
Emotion Sad and regretful
Evidence (words/phrases)
- Linger
- I wish it hadn't gone so fast
- I contemplate the past

Pace   Moderately slow

The lesson concluded with Kathy providing each student with a different poem, for which they were required to use the table as preparation for reading the poem to the class at an appropriate emotional pace in the next lesson.

> **Questions**
> 1 Comment on the approach Kathy used to incorporate numeracy into her teaching.
> 2 Discuss with a partner which aspects of the principles for numeracy task design and implementation are evident in this case study. Which aspects weren't evident?
> 3 Comment on why the missing aspects might not be there, or at least might not be visible.

CONCLUSION

In this chapter, we have examined general principles of task design and a framework for the design and implementation of numeracy tasks as part of planning for how to enhance students' numeracy capabilities across the curriculum. In discussing the framework, approaches to the ways teachers go about identifying, shaping and actualising numeracy tasks in their classrooms were described. Given that there are both demands and opportunities for the incorporation of numeracy tasks across the curriculum, approaches teachers use in different levels of schooling (e.g., primary, secondary) and varying learning areas were also considered. In Chapter 7, we will shift the focus of design and planning from individual lessons and lesson sequences to the whole-school level.

RECOMMENDED READING

Geiger, V., 2016, 'Teachers as designers of effective numeracy tasks', in B. White, M. Chinnappan & S. Trenholm (eds), *Opening Up Mathematics Education Research*, Proceedings of the 39th annual conference of the Mathematics Education Research Group of Australasia, pp. 252–9, Adelaide: MERGA

Goos, M., Geiger, V. & Dole, S., 2013, 'Designing rich numeracy tasks', in C. Margolinas (ed.), *Task Design in Mathematics Education*, Proceedings of ICMI Study 22, July 2014, Oxford, pp. 589–98, Oxford: ICMI, <https://hal.archives-ouvertes.fr/hal-00834054/file/ICMI_STudy_22_proceedings_2013-10.pdf>, retrieved 4 March 2018

Maass, K., Garcia, J., Mousoulides, N. & Wake, G., 2013, 'Designing interdisciplinary tasks in an international design community', in C. Margolinas (ed.), *Task Design in Mathematics Education*, Proceedings of ICMI Study 22, July 2014, Oxford, pp. 367–76, Oxford: ICMI, <https://hal.archives-ouvertes.fr/hal-00834054/file/ICMI_STudy_22_proceedings_2013-10.pdf>, retrieved 4 March 2018

Sullivan, P., Clarke, D. & Clarke, B., 2013, *Teaching with Tasks for Effective Mathematics Learning*, New York: Springer

Sullivan, P. & Yang, Y., 2013, 'Features of task design informing teachers' decisions about goals and pedagogies', in C. Margolinas (ed.), *Task Design in Mathematics Education*, Proceedings of ICMI Study 22, July 2014, Oxford, pp. 529–30, Oxford: ICMI, <https://hal.archives-ouvertes.fr/hal-00834054/file/ICMI_STudy_22_proceedings_2013-10.pdf>, retrieved 4 March 2018

7

Whole-school approaches to numeracy

Planning to embed numeracy across the whole-school curriculum requires more than the efforts of individual teachers. Morony et al. (2004) suggested that such initiatives instead need to involve teams of people, including those with and without mathematical interests and expertise, who work together to understand issues and develop approaches and strategies. They emphasised that 'success with numeracy across the curriculum can only come when numeracy truly becomes *everybody's business*' (p. 4; original emphasis).

This chapter examines some of the challenges of developing a whole-school approach to numeracy in primary and secondary schools. We begin the chapter by revisiting the idea of numeracy across the curriculum and then review Australian research on school leadership in relation to numeracy, to identify challenges in taking a coordinated cross-curricular approach. The remainder of the chapter offers practical strategies for getting started with engaging a whole-school community in numeracy planning, teaching and assessment.

NUMERACY ACROSS THE CURRICULUM... REVISITED

In Chapter 2, we saw how, in the mid-1990s, state and national governments in Australia began to signal increasing interest in improving students' levels of literacy and numeracy, foreshadowing the release of a national numeracy plan that included assessment of students against agreed benchmarks. The national Numeracy Education Strategy Development Conference soon followed, to seek input from the education community on issues surrounding numeracy education (DEETYA 1997). Conference participants, representing education systems, mathematics teaching and educational research organisations, curriculum developers, parent groups and school principals, identified numeracy as involving 'using some mathematics to achieve some purpose in a particular context' (p. 13). Thus, while numeracy is underpinned by mathematics, having knowledge of mathematics is not enough to ensure that students become numerate. An implication of this position is that numeracy is a cross-curricular issue and, like literacy, is therefore 'everyone's business' (p. 12). Conference participants agreed that numeracy needed to be viewed in the same way as literacy—that is, as an essential requirement for learning in different areas of the curriculum.

The Australian Department of Education, Training and Youth Affairs (2000) subsequently developed a numeracy policy for

Australian schools that acknowledged the numeracy demands in all learning areas across the curriculum. The idea of numeracy as a cross-curricular issue was also enshrined in the Australian Curriculum, with numeracy being identified as one of seven general capabilities to be developed within the content of each learning area. These policy initiatives clearly support the need for a coordinated, whole-school approach to numeracy. Yet, the notion of embedding numeracy learning across the curriculum has gained little ground in Australian schools (Carter et al. 2015; Thornton & Hogan 2004).

CHALLENGES TO WHOLE-SCHOOL NUMERACY

The Australian Curriculum provides some assistance in identifying the numeracy demands of different learning areas (see Chapter 4), but it has been left to individual schools to work out how to embed numeracy across the curriculum as a general capability. Although there is a body of research on pedagogical approaches for developing numeracy in different learning areas (see Chapters 5 and 6), few studies have looked at numeracy from a school leadership perspective. One such study was conducted by Carter (2015) to compare the ways in which three secondary schools—one each from the government, Catholic and independent school sectors—interpreted and applied the Australian Curriculum requirement to embed numeracy throughout the curriculum. Carter's study (reported in Carter et al. 2015) investigated the actions of school leaders and teachers in the context of standardised testing of numeracy via NAPLAN. She interviewed principals, curriculum leaders, heads of mathematics and numeracy and teachers of mathematics and numeracy, as well as observing lessons and analysing school documents and emails. While the schools differed from each other in many ways, there were also similarities in their approaches to numeracy.

The government high school had recorded NAPLAN results well below the national minimum standard. This led to an explicit focus on literacy and numeracy, including extensive data analysis, NAPLAN practice tests, appointment of a head of numeracy and introduction of a Year 8 subject labelled 'numeracy' that aimed to remediate mathematical skills. However, the existence of this subject appeared to lead the school's managers and teachers towards a basic skills view of numeracy that did not align with the Australian Curriculum's cross-curricular positioning of numeracy as a general capability to be developed in all learning areas. The Catholic school recorded NAPLAN results similar to like schools, and so literacy and numeracy were not seen to be a problem. Perhaps as a consequence, there was no school-wide approach to embedding numeracy and no individual charged with this responsibility, despite some attempts by the school's curriculum coordinator to encourage teachers in this direction. The independent school recorded NAPLAN results well above the state average, and so literacy and numeracy were not seen to be a problem for most students. The two teachers with leadership in mathematics (head of mathematics and middle school mathematics coordinator) saw numeracy as being separate from mathematics, but neither considered they should play a role in embedding numeracy in non-mathematics subjects. The school's curriculum leader believed that heads of department were not committed to embedding numeracy in their subjects and that individual teachers were reluctant to adopt this approach.

Carter's (2015) research identified the following potential

challenges to embedding numeracy across the curriculum in secondary schools:

- School leaders and teachers did not share the same understanding of the meaning of numeracy.
- There was a lack of clarity in who was responsible for developing numeracy across the curriculum. Heads of mathematics did not see themselves as having this role.
- There was uncertainty as to where numeracy should be taught. Heads of mathematics believed that numeracy should be developed in contexts outside the mathematics classroom, while non-mathematics teachers believed that numeracy involved content that should be learned in mathematics lessons.
- There were no school-wide processes for ensuring that numeracy was being embedded across all subjects.
- Non-mathematics teachers lacked confidence in their own mathematical abilities and lacked knowledge of how to incorporate numeracy into their pedagogy.

RESOURCES FOR WHOLE-SCHOOL NUMERACY

Carter et al. (2015) observed a general lack of commitment to cross-curricular numeracy in their case study schools but acknowledged that schools had received little implementation guidance from curriculum authorities. They suggested that the authorities could prepare classroom-ready units of work and assessment items that embed numeracy in various learning areas and year levels, in order to encourage busy or apprehensive teachers to take the first step in this direction. At the time of writing this book, there are few such resources available to Australian teachers.

> **Review and reflect 7.1**
>
> Carter's (2015) study was conducted in secondary schools, which differ substantially from primary schools in their organisational structures and teacher characteristics. In particular, secondary schools organise learning around subjects that are separately timetabled and taught by teachers who are expected to have subject-matter expertise. How relevant is Carter's list of challenges to embedding numeracy across the primary school curriculum?
>
> 1. With a primary school teacher colleague, create a list of challenges for primary schools in embedding numeracy across the curriculum.
> 2. In your own school, to what extent is there any planning for embedding numeracy across the curriculum?
> 3. Who is responsible for leading numeracy development and curriculum planning in the school?

The scarcity of resources that might support school-wide development of numeracy was revealed by an audit that the authors of this book conducted on behalf of the Queensland College of Teachers (Goos, Geiger, Bennison et al. 2015). We were interested in ways in which existing resources supported teachers' understanding and enactment of numeracy across the curriculum, so we constructed an audit framework to record them. The framework consisted of statements sourced from the numeracy standards for graduate teachers published by the Board of Teacher Registration, Queensland (2005), which describe 'professional knowledge', 'professional practice' and 'professional attributes' in relation to numeracy. The numeracy standards comprise 22 statements, four of which were selected for the audit framework because they refer to understanding (professional knowledge) and enactment (professional practice) of numeracy across the curriculum. The statements are quoted below as they appeared in the audit, for the purposes of which they were preceded by the sentence stem 'How might this resource help teachers to . . .?'

Professional knowledge
PK1: Understand the meaning of numeracy within their curriculum areas.
PK2: Recognise numeracy learning opportunities and demands within curriculum areas.

Professional practice: planning
PPP: Take advantage of numeracy learning opportunities within their curriculum context.

Professional practice: teaching
PPT: Demonstrate effective teaching strategies for integrating numeracy learning within their own curriculum context.
(Goos, Geiger, Bennison et al. 2015, p. 11)

We limited our search for numeracy resources to those that are readily accessible to Australian teachers and endorsed or produced by the authorities responsible for the Australian Curriculum or the Australian Professional Standards for Teachers or by teacher professional associations. Three main sources that we searched were the illustrations of practice that accompany the Australian Professional Standards for Teachers—an online professional development package comprising video clips of classrooms, teacher interviews and discussion questions (AITSL 2017); the government-endorsed repository of digital resources mapped to the Australian Curriculum and available via Scootle (<www.scootle.edu.au>); and teacher professional journals in mathematics and non-mathematics subjects.

The first source of numeracy resources was found to provide little assistance in understanding and enacting numeracy across the curriculum. Only two of the 325 illustrations of practice were related to numeracy, and only one of these (titled 'Embedding mathematics in everything') connected mathematics to non-mathematical contexts—but in the form of extra-curricular activities rather than other school subjects.

For the second source, a search of Scootle using the term numeracy returned 235 resources, almost all of which were related to the teaching of mathematics rather than numeracy across the curriculum. Seventeen numeracy resources were identified, all of which were judged to have the potential to help teachers understand the meaning of numeracy within a particular curriculum area and, if implemented as directed, to help teachers demonstrate effective teaching strategies for integrating numeracy learning in this curriculum context. For example, a unit of work in the science curriculum, on plants, included activities involving measurement of plant growth, development of a scale for a cross-section diagram and the collection and representation of data in tables and graphs.

The third source of numeracy resources was teacher professional journals. A search of issues of 17 journals published in the previous ten years and aimed at teachers of science, English, mathematics, computing, HPE, English as a second language, modern languages, geography, art, history and music, as well as more general journals focusing on early childhood or middle years education, found only 15 articles on the teaching of numeracy across the curriculum. Eleven of these were published in mathematics teacher journals, which are unlikely to be read by teachers of other subjects looking for help in understanding and enacting numeracy in their own curriculum contexts.

Most resources that we found did offer some explanations or examples that could enhance teachers' understanding of the meaning of numeracy in their own curriculum context, and many also provided ready-made activities for integrating numeracy into the teaching of subjects other than mathematics. However, almost none addressed the need for teachers to recognise and take advantage of the numeracy learning demands and opportunities within the subjects they teach as part of their curriculum planning and pedagogical practice.

Review and reflect 7.2

The numeracy resources audit searched for materials that could be used by individual teachers in their classrooms rather than tools for whole-school numeracy planning.

1. Conduct a Google search on whole-school numeracy.
2. What interpretations of numeracy are evident in the sources you find (e.g., numeracy as basic skills remediation, numeracy as primary school mathematics)?
3. Do any sources provide support for embedding numeracy across all subjects in the curriculum?

WHOLE-SCHOOL NUMERACY STRATEGIES

Although this chapter is concerned with whole-school approaches to embedding numeracy, we have learned from our experience in working with teachers in primary and secondary schools that it is important first to build teachers' capacities to identify numeracy demands and opportunities in the subjects they teach. Chapters 4, 5 and 6 provide guidance for individual teachers on how to approach these tasks. Our intention in this section is to present a professional development model that schools could use with small groups of teachers to begin raising awareness of numeracy across the curriculum. The model draws on our previous research (Goos, Geiger & Dole 2014) and ideas developed by Hogan and colleagues (Morony et al. 2004; Thornton & Hogan 2003).

Professional development

The professional development model uses cycles of action research interspersed with one- or two-day workshops, preferably with input from an external expert or team (e.g., consultant, expert teacher, researcher). Table 7.1 shows a typical year-long professional development plan, which we have implemented with volunteer groups of 20 teachers from up to ten schools. Although smaller groups would work as well, it is important to include teachers of different subject areas (for secondary schools) and year levels (for primary schools). A secondary school with its feeder primary schools is an effective grouping for such a project.

The first workshop elicits participants' own interpretations of numeracy—for example, by using the task from Review and

Table 7.1 Professional development plan for whole-school numeracy

Months	Activities	Details
March	Professional development workshop	• Elicit interpretations of numeracy • Introduce numeracy model • Try out numeracy teaching strategies and tasks • Plan for Action Research Cycle 1
March–June	Action Research Cycle 1	Design and teach a lesson sequence
June	School visits	• Lesson observations • Interviews with teachers and students • Collection of teaching materials and curriculum planning documents
July	Professional development workshop	• Evaluate implementation • Share teaching resources and strategies • Plan for Action Research Cycle 2
July–October	Action Research Cycle 2	Design and teach a unit of work
October	School visits	As for school visits in June
November	Professional development workshop	• Evaluate implementation • Reflect on professional learning

reflect 1.1—before introducing the model of numeracy to be adopted for classroom use (such as the 21st Century Numeracy Model presented in Chapter 3). Teachers are then engaged in a range of numeracy-embedded tasks offered by the workshop facilitators

(see Figure 7.1 for an example), and they are supported to analyse the features of these tasks through the lens of the numeracy model. In the final session of the first workshop, teachers begin to plan a lesson sequence that they will teach in the coming term.

> **Review and reflect 7.3**
>
> 1 Design a numeracy investigation based on the example in Figure 7.1. Use the 21st Century Numeracy Model from Chapter 3 and design principles from Chapter 6 to assist you with this task.
> 2 Share your investigation with a partner and exchange feedback on how it could be improved.
>
>
> - Is a Barbie doll a realistic representation of human proportions?
> - What would it look like if it were scaled up to human height?
>
> Figure 7.1 Sample numeracy investigation
> Source: Shutterstock.
>
> The manufacturers of the doll Barbie have recently added new products to the range, with a much wider and more realistic variety of physical features.
>
> 3 Read the article reporting on this development (Hart 2016).
> 4 If possible, obtain several different versions of Barbie and compare their physical proportions in relation to adult females.

In the second workshop, teachers give short presentations to their peers to share their experiences of designing and teaching a sequence of lessons that respond to the numeracy demands of a specific subject in the curriculum. The sharing of more and less successful designs, with reflections on what they observed in their students, is a powerful learning experience for all teachers and leads to more effective planning of an entire unit of work in the next action research cycle.

A similar approach is taken in the final workshop, with the emphasis shifting to evaluation of the whole project and reflection on what the teachers have learned. This is facilitated by giving each teacher a copy of the numeracy model and asking them to annotate it to represent their trajectory through the model as they became more familiar with each of its dimensions. (See Figure 7.2 for an example.)

We have found that most teachers begin with the desire to improve their students' mathematical knowledge or dispositions and then move on to explore tools and contexts. Typically, by the end of the year, only a small proportion of teachers has engaged with the critical orientation element of the model, as this seems to be the most challenging aspect to understand and implement. Additional activities may be needed to support teachers in understanding what it means to be critical, such as the example provided in Review and reflect 7.4.

Hogan and colleagues describe a similar professional development model that they refer to as the Numeracy Research Circle (see Hogan 2002). In this approach, the first action research cycle focuses on identifying numeracy moments, 'incidents in which students encountered mathematical ideas in other contexts' (Thornton & Hogan 2003, p. 122). One example is presented below:

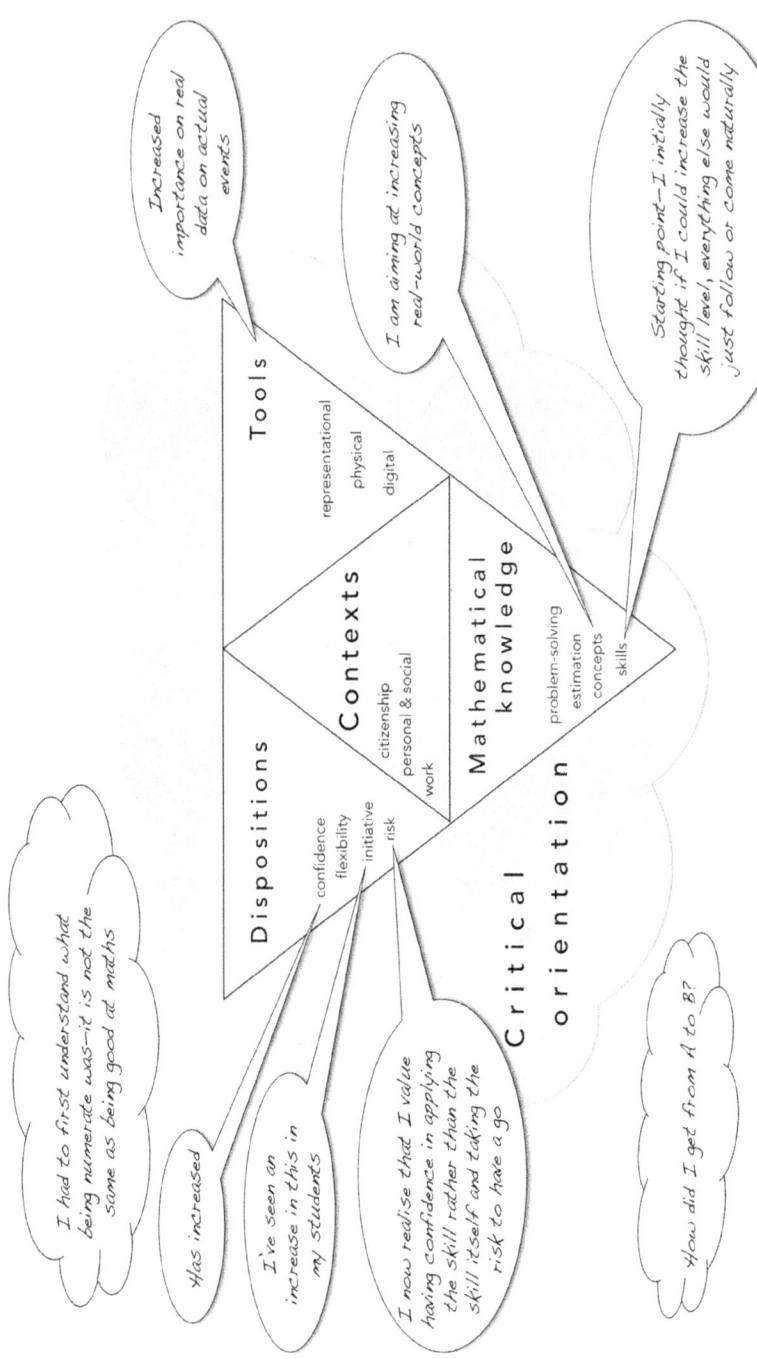

Figure 7.2 One teacher's trajectory through the 21st Century Numeracy Model

Source: Geiger, Goos & Dole 2011.

> **Review and reflect 7.4**
>
> So many cars have been advertised in the *Daily Planet* classifieds that, lined up, they could circle the moon.
>
> Analyse the advertising claim above and compare your views with those of a partner. Collaboratively design a set of questions to help your students take a critical orientation to this information. A critical orientation can involve the following:
>
> - making a judgement
> - making a decision
> - supporting an argument
> - disputing a claim.

The volume experiment

Viktor observed the discrepancy between students' capacity to measure volume accurately in science and their ability to apply mathematical formulas for volume in mathematics. Despite measuring the same objects by displacement of water and using a ruler, students seemed unaware that obtaining divergent answers created a problem. They seemed unaware of issues such as appropriate levels of accuracy. (p. 123)

Numeracy audit

Morony et al. (2004) describe a numeracy audit as a means by which teachers and school leaders can collect information to begin planning for a whole-school approach to numeracy development

> **Review and reflect 7.5**
>
> Work with a colleague in your school to document numeracy moments. This means recording, in as much detail as you can, the circumstances in which your students encounter mathematical ideas in the subjects you teach, any problems they have in understanding the mathematics or the context, what you did as a result of these observations and what the students did next (Thornton & Hogan 2003, p. 122).

(see also Hogan 2000b). The key steps in conducting a numeracy audit are summarised in Table 7.2, together with suggestions for how to collect evidence to inform planning of a whole-school approach.

The numeracy audit can be conducted for the whole school or a smaller subset of the school, such as a single learning area (e.g., HPE), a single year group (e.g., Year 9) or a phase of schooling (e.g., lower primary).

In Case study 7.1 we present details of primary and secondary schools that we have worked with to develop whole-school curriculum leadership.

Table 7.2 Numeracy audit approach

Steps	Sources of evidence
1 Collect and organise background information on the school, its students and staff	• School records • School website • My School website (<www.myschool.edu.au>)
2 Identify staff perceptions of numeracy and confidence in recognising numeracy demands and opportunities in the subjects they teach	• Workshops or interviews using Review and reflect 1.1 • Self-assessment of confidence in numeracy teaching using Tables 1.2 and 1.3
3 Identify student perceptions of numeracy	• Focus group interviews • Surveys
4 Examine school and curriculum documents for descriptions of numeracy	• Australian Curriculum (learning areas and general capabilities) • Subject department or year level planning documents • Audit of numeracy demands (see Chapter 4)
5 Review teachers' own curriculum planning documents for references to numeracy	• Teachers' personal planning documents • Audit of numeracy demands (see Chapter 4)
6 Collect numeracy examples from the classroom	• Collect and share with colleagues one significant numeracy example and analyse it using the numeracy model introduced in Chapter 3 • Observe and record numeracy moments over a four-week period • Shadow a student across different learning areas over a one-week period
7 Analyse and display data, discuss and identify issues arising from analysis, plan future action, and continue with action research cycle	• Further surveys and interviews with teachers and students • Photos and videos of lessons • Student work samples • Communication with parents via newsletters and emails

Source: Adapted from Morony et al. 2004.

Review and reflect 7.6

1. Revisit the numeracy self-assessment survey that you completed in Table 1.2 or 1.3. In which areas do you now feel most confident and least confident?
2. Compare your results with those of a colleague teaching the same subject, year level or both.
3. Give the Review and reflect 1.1 task to a group of students you teach and ask them to respond to the questions it contains.
   - Conduct a discussion with the group that asks them to compare and elaborate on their responses and create an agreed definition of what they think numeracy is.
   - Ask the students to interview their parents to find out their perceptions of numeracy.

Case study 7.1 Whole-school numeracy in primary and secondary schools

Primary school 1

The following narrative was written by one of the authors of this book who worked as a consultant with Barbara, who previously participated in a project that introduced her to the 21st Century Numeracy Model. Barbara taught in a small school with one class per year level. All teachers had made a commitment to embed numeracy across the whole curriculum, and Barbara was the action learning team leader who was responsible for leading this initiative.

The school had started using a new mathematics textbook, but Barbara was also asking teachers to plan more explicitly for numeracy across the curriculum using a proforma she provided. Teachers had to give their Term 2 plans to Barbara for review; she found the range of topics quite limited. She organised a day when each teacher met with her individually to pull out the links to the textbook and the numeracy model in their plans. Teachers responded positively to these meetings. Teachers subsequently implemented their plans and were ready to start planning with Barbara for Term 3. But Barbara was unsure of how teachers were implementing these plans in their classrooms. The consultant suggested that visiting other teachers' classrooms might be worthwhile, but teachers were not accustomed to this approach. Perhaps a better suggestion was for Barbara to start by inviting other teachers to observe her own class, as this may be less threatening.

Barbara was thinking about how to structure a whole-day in-service for staff. Her main challenges were in building teachers' confidence in planning and helping them to see that they might only need to make small changes in their practice. So far, she hadn't placed the numeracy model at the forefront of her work with staff, because she was already familiar with it. She was therefore thinking of revisiting the model at the in-service day, making posters of the model for each teacher, having the model available during collaborative planning sessions and explicitly listing the elements of the model in planning documents so that embedding of numeracy could be documented and audited. At the in-service day, the model could be used to analyse units taught last term and planned for the next term. This would highlight elements of the model that received emphasis and those that could be emphasised more.

Barbara mentioned that she still had trouble incorporating a critical perspective into her lessons. But she described a lesson from her current Year 6 history unit that illustrated how to take advantage of serendipitous moments. The unit was about Australian government, and students were to imagine they are prime minister and complete a set of related tasks. One was about the characteristics of Australia's population. Barbara had prepared questions for students to investigate, but when she overheard one student ask a peer 'Are Aboriginal people counted in the census?' she decided to ask students to write down all the questions that they wanted to ask. This exercise yielded a rich set of student-generated questions for developing a critical numeracy perspective on historical inquiry. The consultant suggested that she might try to recreate this incident at the in-service day and ask teachers if they could write their own questions as a way of developing their own critical orientation.

Question

Discuss with a partner the strategies that Barbara was using to engage other teachers with the idea of embedding numeracy across the curriculum. What additional strategies might she try?

Primary school 2

Three teachers in this school formed a team to participate in a numeracy across the curriculum leadership project led by some of the authors of this book. Their goals were to build the capacity of other teachers to embed numeracy across the curriculum, in order to 'widen the circle in numeracy' within the school.

The numeracy team began by designing and administering a survey to identify staff perceptions of numeracy and their confidence in numeracy planning and teaching. The survey items are shown in Table 7.3. From the results, the team concluded that their colleagues believed in the importance of numeracy as an essential skill for life and learning but lacked confidence in planning for numeracy for their classes and were possibly unaware of the numeracy demands in curriculum areas other than mathematics.

Table 7.3 Survey of teacher perceptions of numeracy

Statement	Rating 5-1
I believe numeracy is an essential skill for life and learning	
I enjoy teaching mathematics	
I am confident in planning for numeracy for my class	
I am aware of and act upon the numeracy demands across the curriculum	
I embrace numeracy opportunities as they arise	
I provide a range of real-life activities for the teaching of numeracy	
My current teaching practices develop students' numeracy capability	
I would rate my current understanding of numeracy in curriculum areas other than mathematics as...	

Note: Ratings: 5 very confident, 4 confident, 3 unsure, 2 unconfident, 1 very unconfident.

The leader of the numeracy team devised a numeracy filter that allowed teachers to analyse the numeracy demands of any curriculum area in terms of the organising elements for numeracy as a general capability in the Australian Curriculum. An example is shown in Table 7.4. The team developed the timeline shown in Table 7.5 for trialling the numeracy filter and sharing their work with other teachers in the school. When they reflected on how to sustain this initiative into the following year, they suggested that they would need to:

- increase the size of the numeracy team to four or five teachers in order to avoid staleness or fatigue
- continue to model use of the numeracy filter to plan units of work
- work with other teachers to embed numeracy into science and history planning
- lead their year level cohorts in numeracy planning in other curriculum areas.

Table 7.4 Use of a numeracy filter to plan for embedding numeracy into the history curriculum

Elements of numeracy as a general capability	Examples
Calculating and estimating	• Using number to order events by date • Recognising that a 100s chart can represent four generations • Rounding a population to the nearest 1000
Recognising and using patterns and relationships	• Looking for patterns and relationships in immigration trends over time
Using fractions, decimals, percentages, ratios and rates	• Using data to calculate percentages of votes for and against federation • Finding the percentage of Australians born overseas

Elements of numeracy as a general capability	Examples
Using spatial reasoning	• Creating a three-dimensional model of a building from the past • Using maps to explain routes followed by explorers or patterns of development in the Australian colonies
Interpreting and drawing conclusions from statistical information	• Organising and displaying data about different groups of people on the First Fleet • Interpreting life expectancy data for Aboriginal people
Using measurement	• Using the language of time to describe duration of events • Constructing annotated timelines for key people and events • Using measurements from maps, plans and other sources to describe historical buildings and the layouts of settlements

Questions

1. Compare the strategies used in the two primary schools and comment on the similarities and differences between them.
2. Evaluate the strategies used by both schools in light of the numeracy audit approach summarised in Table 7.2.
3. Make some recommendations for how both schools might sustain their numeracy development work over the next two years.
4. Compare your recommendations with those of a colleague.

Table 7.5 Plan for initial development of whole-school numeracy approach

Months	Activity
April	Staff information session and pre-survey
May	Team leader develops numeracy filter
June	Team leader plans and teaches trial unit designed with numeracy filter
August	Report on progress at staff meeting
September	Team leader works with other team members to use numeracy filter in planning for Term 3
October	Expand use of numeracy filter to planning with four other teachers (outside the team)
November	Post-survey participating teachers and share project summary with staff

Secondary school

We have seen from Carter's (2015) research, discussed earlier in this chapter, that there are significant challenges in embedding numeracy across the curriculum in secondary schools. Our secondary schools case study consists of a set of video resources that illustrate how some secondary schools are addressing these challenges. Visit the University of Queensland (2018) page at the Filmpond website and watch the following short videos on embedding numeracy across the curriculum in a secondary school: 'Year 9 Design and Technology', 'Year 9 German', 'Year 10 Health' and 'Establishing a Numeracy Committee'.

Questions

1 How might these resources help secondary school leaders establish a supportive and challenging learning environment that values numeracy learning?
2 How could you develop and communicate informed perspectives on numeracy within and beyond the secondary school?

CONCLUSION

Since the literacy across the curriculum movement of the 1970s, it has become widely accepted that all teachers are responsible for contributing to their students' literacy development within the subjects they teach. Thus, for example, scientific literacy should be developed in the context of school science, and mathematical literacy within school mathematics. Numeracy needs to be viewed as a cross-curricular concern in much the same way as literacy. However, cross-curricular approaches to numeracy have received little support in practice, possibly because of the widespread disaffection with mathematics experienced by so many people—including teachers who are not mathematics specialists.

In this chapter, we have argued for the importance of school leaders and teachers developing a whole-school approach to embedding numeracy across the curriculum. This approach is not without challenges. We suspect that the most productive way forward might be to focus on numeracy as a key enabler of learning in different areas of the curriculum rather than using subject learning as a means of improving students' numeracy performance.

RECOMMENDED READING

Carter, M., Klenowski, V. & Chalmers, C., 2015, 'Challenges in embedding numeracy throughout the curriculum in three Queensland secondary schools', *Australian Educational Researcher*, vol. 42, pp. 595–611

Geiger, V., Goos, M. & Dole, S., 2011, Teacher professional learning in numeracy: Trajectories through a model for numeracy in the 21st century, in J. Clark, B. Kissane, J. Mousley, T. Spencer & S. Thornton (eds), *Mathematics: Traditions and (new) practices*, Proceedings of the 23rd biennial conference of the Australian Association of Mathematics Teachers and the 34th annual conference of the Mathematics Education Research Group of Australasia, pp. 297–305, Adelaide: AAMT & MERGA

Goos, M., Geiger, V., Bennison, A. & Roberts, J., 2015, *Numeracy Teaching Across the Curriculum in Queensland: Resources for teachers; Final report*, <http://qct.edu.au/pdf/Numeracy_Teaching_Across_Curriculum_QLD.pdf>, retrieved 17 August 2017

Thornton, S. & Hogan, J., 2003, 'Numeracy across the curriculum: Demands and opportunities', paper presented at the annual conference of the Australian Curriculum Studies Association, Adelaide, 28–30 September, <www.acsa.edu.au/pages/images/thornton_-_numeracy_across_the_curriculum.pdf>, retrieved 18 August 2017

8

Assessing numeracy learning

In this chapter, we explore how numeracy is assessed across Australia and how the results are interpreted. You will learn that what purports to be an assessment of numeracy is, in reality, something else. The appropriate and inappropriate interpretations of testing regimes are also considered.

Numeracy is one of seven general capabilities that teachers, across all disciplines and at all year levels, are expected to foster and develop in their students. According to the Australian Curriculum, Assessment and Reporting Authority, 'Numeracy encompasses the knowledge, skills, behaviours and dispositions that students need to use mathematics in a wide range of situations. It involves students recognising and understanding the role of mathematics in the world and having the dispositions and capacities to use mathematical knowledge and skills purposefully' (ACARA 2018a). As described in the Foundation to Year 10 Australian Curriculum, six key ideas comprise numeracy: 'Using spatial reasoning', 'Interpreting statistical information', 'Using measurement', 'Estimating and calculating with whole numbers', 'Recognising and using patterns and relationships', and 'Using fractions, decimals, percentages, ratios and rates' (ACARA 2018e).

The Australian Curriculum's definition of numeracy and the definition encompassed by the 21st Century Numeracy Model frame all discussion in this book. As discussed earlier in this book in relation to the 21st Century Numeracy Model, context is central to the tasks and activities that foster students' numeracy capabilities. To refresh your memory, you may want to turn back to the schematic for the 21st Century Numeracy Model in Figure 3.1. How numeracy is assessed in Australia, how the results are interpreted and how classroom teachers in all subject disciplines can gauge students' numeracy competencies are the foci of this chapter.

> ### Review and reflect 8.1
>
> Before reading further in this chapter about how numeracy is assessed in Australia, reflect on your current knowledge of the following issues:
>
> 1 How is numeracy tested in Australia? Who is tested? What is learned from this testing?
> 2 Which international testing regimes does Australia participate in? Who is tested? What are the impacts of this testing?

NATIONAL AND INTERNATIONAL ASSESSMENTS OF NUMERACY

In this section, the various national and international numeracy assessment programs in which Australian school students and adults (aged 16–65) participate are discussed. We begin with three assessment regimes in which school-aged students participate: NAPLAN, PISA and the Trends in International Mathematics and Science Study (TIMSS). These are followed by the Literacy and Numeracy Test for Initial Teacher Education Students (LANTITE), which Australian pre-service teachers must pass, and finally, PIAAC, which is completed by a cross-sample of the Australian adult population.

It is important to recognise that, both nationally and internationally, terminology use with respect to numeracy varies. The Australian Curriculum and PIAAC definitions of numeracy, together with PISA's definition of mathematical literacy, are similar and consistent with the definition of numeracy in the 21st Century Numeracy Model that has been adopted in this book. NAPLAN's implied definition (no exact definition is provided) is different from the others. In NAPLAN, the numeracy test is essentially assessing mathematical knowledge against the content strands at the various year levels that are outlined in the Australian Curriculum: Mathematics.

NAPLAN

The term numeracy has been misused in the Australian context. Since 2008, all Australian students in Years 3, 5, 7 and 9 have been required to complete NAPLAN (National Assessment Program Literacy and Numeracy) tests. Interestingly, no definition of numeracy can be found in the National Assessment Program documentation. We are told that the Australian Curriculum: Mathematics is used as the base reference for the numeracy tests, which 'assess the proficiency strands of understanding, fluency, problem-solving and reasoning across the three content strands of mathematics: number and algebra; measurement and geometry; and statistics and probability' (NAP 2016c). Effectively, as we argued in Chapter 2, NAPLAN numeracy is a measure of mathematics achievement relative to the Australian Curriculum and what is expected that students have encountered in their mathematics learning at school.

It has also become common to refer to the mathematics curricula and mathematics learning in the early years and primary levels across Australia as numeracy learning. In primary contexts, we have had numeracy leaders and numeracy coaches, but in reality their work has only focused on mathematics learning and effective teaching approaches at the early years and primary levels. At the secondary level, the term numeracy is not used to equate to the subject mathematics that is taught. Unfortunately, some educators continue to talk about numeracy at the early years and primary levels when they mean mathematics, and this can be particularly confusing for pre-service teachers.

As discussed above, NAPLAN is a national testing regime and, although termed numeracy, it is the mathematics performance of students in relation to the Australian Curriculum: Mathematics that is being tested. One of the positive aspects of the NAPLAN numeracy test, compared to tests in earlier times, is that many of the questions are set in context (see Figure 8.1).

Review and reflect 8.2

Figure 8.1 shows one question taken from the *Numeracy: Calculator allowed; Year 7 example test* (NAP & ACARA 2017). You might want to refer back to the diagram of the 21st Century Numeracy Model (Figure 3.1) to reflect on the questions below.

2 For 3 days, Bella made a tally of the birds she saw in a park. This table shows her results.

Type of bird	Monday	Tuesday	Wednesday
Kookaburra	I I		I
Magpie	I	I I	I
Galah	I I I	I I	
Rosella		I I I I	I I

Which column on the graph below shows the total number of Galahs?

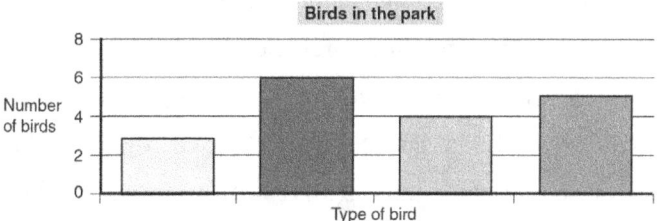

Figure 8.1 NAPLAN numeracy test sample item
Source: NAP & ACARA 2017.

1 Is this an authentic context? Why?
2 What might make the context relevant to a particular class of students?
3 What might be the teacher's aim in seeking (or providing) observational data such as those given here?
4 What numeracy skills might the teacher be aiming to develop?

> 5 Which tools, if any, might students need to use in order to develop these skills?
> 6 What might constitute a critical dimension to the task?

What can be learned from NAPLAN results?

Many data are generated from NAPLAN testing. School results are made available on the My School website (<www.myschool.edu.au>). Schools receive information on the performance on the tests of each of their students, and parents also receive the results for their children (see NAP 2016g). National reports are prepared (see NAP 2016b; for details of what is included in the reports, see ACARA 2016).

It is very important that schools, and the teachers who work in them, can appropriately interpret the NAPLAN results (see NAP 2016a). The interpretation requires knowledge of what information is included in the scales (NAP 2016d), standards (NAP 2016f) and score equivalence tables (NAP 2016e) provided. As well, the pertinent personal numeracy skills associated with graphical and statistical interpretation are needed.

When appropriately interpreted, the data enable schools to identify strengths and weaknesses in their teaching programs in the areas tested in NAPLAN: literacy and numeracy (in reality, mathematics, as discussed earlier). School- and grade-level priorities can be set addressing shortcomings and for the provision of pertinent professional development. Without the appropriate numeracy skills, school leadership teams and individual teachers are not able to capitalise on the opportunities to address weaknesses in their teaching programs that will, in the longer term, benefit their students and their learning.

> **Review and reflect 8.3**
>
> 1 Go to the My School website (<www.myschool.edu.au>). Use the 'Find a school' search facility to select a school.
> 2 School profile data are provided first: school facts, school staff information, links to the school website, student background information and information about the students enrolled in the school. You can click on a link to find information about the school's finances.
>   - Why do you think it might be important to have information about the staff, student background, enrolled students profile and school finances?
>   - How might these factors impact on the interpretation of the NAPLAN performance data?
> 3 When you click on the NAPLAN link, you will find five further links: 'Graphs', 'Numbers', 'Bands', 'Student gain' and 'Similar schools'. You need to explore each of these. When you click on 'Graphs', you will see that you have the option to select:
>   - a year level and whether you wish to look at the literacy ('Reading', 'Writing', 'Narrative writing', 'Spelling' and 'Grammar and punctuation') or the numeracy results.
>   - You also can choose between 'Selected school', 'Schools with similar students' or 'All Australian schools'. (Each of these options is worth looking at.)
>   - Finally, you can choose to look at the results in 'Bands' and 'Scores'. (Again, each of these options is worth examining.)

If you choose to view a school's results in 'Bands', scroll down the page to see additional information about the students participating in the particular test: the percentages of who participated and who were assessed, exempted, absent and withdrawn. (Note, for example, that the percentages work as follows: assessed + exempt = participated). There is also information provided on how to interpret the symbols used on the graphs.

It is very important to note that the NAPLAN results data are provided for a number of years. These data allow schools and their teachers to look at trends over time and thus identify how performance may have changed over time. Trying to find reasons and explanations for the changes is one of the major challenges; this information is left to schools to work through. It is far too simple to only consider the students as the source of any variations.

4 Look closely at the selected year level in your chosen school and any variation in performance over time. What factors do you believe should be considered when trying to find explanations for any variations you might observe?

PISA

Every three years, schools are randomly selected to participate in PISA (Programme of International Student Assessment), run by the OECD; that is, not all Australian children complete the assessment. In the selected schools, 15-year-old students participate in the assessment, whether or not they are studying mathematics. One of the PISA tests is a test of mathematical literacy, which is defined as 'an individual's capacity to formulate, employ, and interpret mathematics in a variety of contexts. It includes reasoning mathematically and using mathematical concepts, procedures, facts and tools to describe, explain and predict phenomena. It assists individuals to recognise the role that mathematics plays in the world and to make the well-founded judgements and decisions needed by constructive, engaged and reflective citizens' (OECD 2016, p. 65).

The PISA mathematical literacy test is not curriculum based; rather, it is essentially a test of mathematical skills that are set in contemporary everyday contexts. It is clear that the definition of mathematical literacy adopted in PISA resonates closely with the definition of numeracy found in the Australian Curriculum (see above).

Since PISA began in 2003, there has been a steady decline in Australian students' mathematical literacy performance. The mean scores are illustrated in Figure 8.2 together with the OECD average scores.

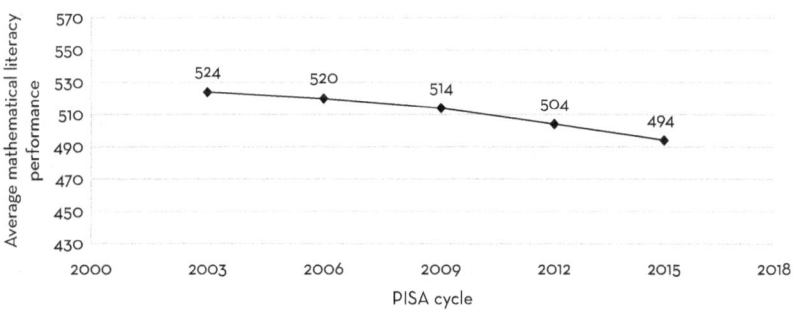

Figure 8.2 Australian students' performance on PISA mathematical literacy over time

Source: Thomson et al. 2017.

The reports of the triennial PISA results from Australia make for interesting reading. The reports can be found on the Australian Council for Educational Research website (see ACER 2018b). Results for mathematical literacy (as well as scientific literacy and reading literacy) are reported by state, gender, location and indigeneity. Comparisons are also drawn with the results from other participating countries. Student attitude data—that is, students' dispositions towards learning mathematics—are also gathered in PISA and found in these reports. PISA data are not reported at the school level, nor do students receive individual results.

PISA results attract the attention of policy-makers and the media. It is disappointing that many educational stakeholders and politicians, as well as journalists, sometimes inappropriately interpret and equate achievements on NAPLAN or PISA, or both, as measures of mathematical knowledge and understanding. Often, too, the relative attainments of schools on NAPLAN are compared or the 'mathematical health' of the nation is discussed with respect to PISA results. The NAPLAN and PISA results are often considered to be measures of the success, or otherwise, of the intended mathematics curriculum and of the teaching of mathematics. If students perform less well than expected—unrealistically, improvement is the common expectation—teachers are frequently made into scapegoats.

As PISA is an international assessment regime, it can be very challenging for the test developers to devise authentic, real-world contexts that will be in the everyday experiences of students from all walks of life. A sample item from PISA is shown in Figure 8.3.

HELEN THE CYCLIST

Helen has just got a new bike. It has a speedometer which sits on the handlebar.

The speedometer can tell Helen the distance she travels and her average speed for a trip.

Question 1: HELEN THE CYCLIST

On one trip, Helen rode 4 km in the first 10 minutes and then 2 km in the next 5 minutes.

Which one of the following statements is correct?

A   Helen's average speed was greater in the first 10 minutes than in the next 5 minutes.
B   Helen's average speed was the same in the first 10 minutes and in the next 5 minutes.
C   Helen's average speed was less in the first 10 minutes than in the next 5 minutes.
D   It is not possible to tell anything about Helen's average speed from the information given.

Question 2: HELEN THE CYCLIST

Helen rode 6 km to her aunt's house. Her speedometer showed that she had averaged 18 km/h for the whole trip.

Which one of the following statements is correct?

A   It took Helen 20 minutes to get to her aunt's house.
B   It took Helen 30 minutes to get to her aunt's house.
C   It took Helen 3 hours to get to her aunt's house.
D   It is not possible to tell how long it took Helen to get to her aunt's house.

Question 3: HELEN THE CYCLIST

Helen rode her bike from home to the river, which is 4 km away. It took her 9 minutes. She rode home using a shorter route of 3 km. This only took her 6 minutes.

What was Helen's average speed, in km/h, for the trip to the river and back?

Average speed for the trip: .... km/h

Figure 8.3 Question from 2012 released PISA items
Source: OECD 2012b.

> **Review and reflect 8.4**
>
> 1 Look at Figure 8.3. Make a comment on the context and the extent to which it might be familiar to the everyday experiences of Australian children.
> 2 To what extent might it be familiar to the everyday experiences of children in other parts of the world? Explain your responses.

TIMSS

The International Association for the Evaluation of Educational Achievement (or IEA) conducts TIMSS (Trends in International Mathematics and Science Study) every four years, to assess Year 4 and Year 8 students' mathematics and science attainments. It is clearly articulated that TIMSS is an internationally agreed curriculum-based testing regime, not an assessment of numeracy capabilities. Australian students participate in the study. (To read more about TIMSS, see International Association for the Evaluation of Educational Achievement 2018; for Australian students' performance, see ACER 2018c, where you can find the four-yearly reports.)

LANTITE

The implementation of LANTITE (Literacy and Numeracy Test for Initial Teacher Education Students) was a recommendation of a report by the Teacher Education Ministerial Advisory Group (2014). All Australian education ministers agreed to introduce the test to provide a nationally consistent way to measure the pre-existing requirement that prior to graduation all initial teacher education students possess personal literacy and numeracy skills equivalent to the top 30 per cent of the adult population. It is noteworthy that the introduction of the test coincided with increased concerns about school students' perceived poor, or lower than expected, performance levels over time in PISA, NAPLAN and TIMSS.

The test definition of numeracy is consistent with that found in the Australian Curriculum and with PISA's definition of mathematical literacy and, as you will see below, is very closely aligned to the definition of numeracy used in PIAAC. Thus, the test is not an assessment of an individual's mathematical knowledge but an authentic appraisal of personal numeracy skills in contexts relevant to school education (ACER 2017b).

PIAAC

The OECD's (2018b) PIAAC (Programme for the International Assessment of Adult Competencies) is an international assessment program of adults' (aged 16 to 65) numeracy capabilities; their capabilities in literacy and in problem-solving in technology-rich environments are also tested. Within PIAAC, numeracy is defined as 'the ability to access, use, interpret and communicate mathematical information and ideas in order to engage in and manage the mathematical demands of a range of situations in adult life' (OECD 2012a, p. 33). The definition is consistent with that of mathematical literacy used in PISA, including the focus on using mathematics in meaningful ways in everyday life, and with the

> **Review and reflect 8.5**
>
> Figure 8.4 shows a sample question for the personal numeracy test in LANTITE. Can you work out what Alex's result is?
>
> SCIENCE RESULT
>
> This table shows the overall achievement required for different awards in a tertiary science subject.
>
> | Award | Achievement |
> | --- | --- |
> | High Distinction | 80% and over |
> | Distinction | 70%–79% |
> | Credit | 60%–69% |
> | Satisfactory | 50%–59% |
> | Unsatisfactory | below 50% |
>
> The science subject has three assessment tasks. Each task is weighted as follows:
>
> Assessment Task 1: weight 60%
> Assessment Task 2: weight 30%
> Assessment Task 3: weight 10%
>
> Alex's result for each task was:
>
> Assessment Task 1: 70%
> Assessment Task 2: 80%
> Assessment Task 3: 90%
>
> What is Alex's award for science?
>
> A   High Distinction
> B   Distinction
> C   Credit
> D   Satisfactory
>
> **Figure 8.4** Sample LANTITE item
> Source: ACER 2018a.

LANTITE definition of numeracy, for which the everyday and school and workplace contexts requiring the use of mathematical skills are emphasised. As well, the PIAAC numeracy definition resonates with the definition of general capability in numeracy in the Australian Curriculum.

The PIAAC (2013–16) numeracy results for participating countries are shown in Figure 8.5. The overall OECD mean score for PIAAC numeracy was 263; Australian participants' mean score was 268, slightly above the OECD average. Somewhat disturbing was the finding that 'adults in Australia show above-average proficiency in literacy and problem solving in technology-rich environments, however it [sic] only shows around the average proficiency in numeracy compared with adults in the other countries participating in the survey' (OECD 2018c, p. 2).

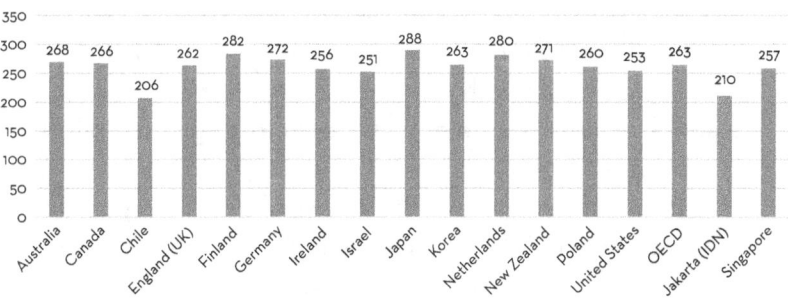

Figure 8.5 PIAAC numeracy performance by country, 2013–16
Source: OECD 2018a.

ASSESSING CLASSROOM NUMERACY DEVELOPMENT

In the next sections, we examine and critique how students' numeracy capabilities are, and can be, assessed in Australian classrooms and how classroom teachers across all discipline areas and at all year levels can gauge their students' numeracy skill levels. We also examine other dimensions of numeracy encompassed by the 21st Century Numeracy Model—tools, dispositions and critical orientation—that should be considered. We consider why it is

> **Review and reflect 8.6**
>
> Look at the data in Figure 8.5. Japan (288) and Finland (282) were the two top-performing countries in PIAAC. Chile (206) and Jakarta (in Indonesia, 210) were the two lowest performing.
>
> 1. Based on what you know of the four countries mentioned above, what factors do you think may have contributed to the superior numeracy performance of adults in Japan and Finland? And what might have been the contributing factors to the poor numeracy attainments of adults in Chile and Jakarta?
> 2. How does the numeracy attainment of Australian adults compare with those of adults in similar English-speaking countries (e.g., New Zealand, England, United States)?
> 3. In terms of current trends in Australia with respect to numeracy, what do you predict might be the outcome in the next PIAAC round (in approximately 2020), particularly among young adults aged 16 to 20, for example?

critical that all teachers have adequate and appropriate numeracy skills to meet the expectations of the Australian Curriculum, as well as the Australian Institute for Teaching and School Leadership's Australian Professional Standards for Teachers.

Australian Professional Standards for Teachers

Earlier in this chapter, we discussed the LANTITE testing of numeracy as a pre-requisite for teacher registration in Australia. LANTITE emerged as a consequence of concerns about teacher quality and the Australian Institute for Teaching and School Leadership's responses to these concerns. As we discussed in Chapter 2, the institute developed the Australian Professional Standards for Teachers at all levels within the profession: graduate, proficient, highly accomplished and leading teachers. Along with the requirement of Standard 2.5, 'Literacy and numeracy strategies: Know and understand literacy and numeracy teaching strategies and their application in teaching areas', which we discussed in the earlier chapter, the graduate standards related to teachers' numeracy capabilities include, in Standard 5.4, 'Interpret student data: Demonstrate the capacity to interpret student assessment data to evaluate student learning and modify teaching practice' (AITSL 2017). For teacher education programs to be accredited, they need to demonstrate that those graduating from their courses meet these and all the other institute standards. Courses on numeracy teaching strategies and their application in teaching areas can also be structured in such a way as to support students to succeed in LANTITE.

Australian Curriculum

Complementing the Australian Institute for Teaching and School Leadership's professional teaching standards is the expectation of the Australian Curriculum that teachers at all year levels (Foundation to Year 10) develop students' numeracy skills across the curriculum—that is, in all the disciplines that they teach. This requires careful consideration by teachers, who need to recognise when numeracy opportunities arise and capitalise on them. There is no expectation in the Australian Curriculum that teachers assess numeracy capabilities across the various subject domains. However, when developing classroom tasks, teachers can, as appropriate, consider including aspects that enable students to demonstrate their numeracy competencies. It is important that teachers recognise that numeracy can promote learning in various subject domains; that is, numeracy skills are integral to learning in specific content areas of the full gamut of school subjects. Case studies 8.1 and 8.2 give examples of tasks in subjects other than mathematics that provide opportunities for numeracy development.

Case study 8.1 A history, health, mathematics or statistics task

Figure 8.6 Florence Nightingale: the lady of the lamp

Florence Nightingale (Figure 8.6) is well known for her nursing during the Crimean War (1853-56), but what do you know of her contributions to the field of statistics and to improvements in hospital care for patients? Watch Hans

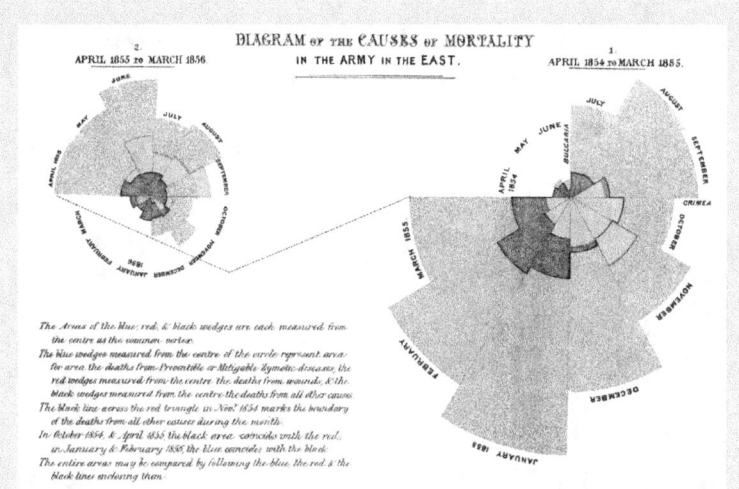

Figure 8.7 Florence Nightingale's polar-area diagram (polar graph)

Rosling's (2011) account of Florence Nightingale's major contribution to the human condition.

Figure 8.7 shows a copy of Florence Nightingale's persuasive graphs related to army deaths that she recorded in a field hospital during the Crimean War. When she returned to England, she used the polar graphs she had devised to mount a strong case that hygiene in hospitals required serious attention. Her argument was successful, and we are all the beneficiaries today.

Questions

1 Based on these polar graphs, explain the case that Nightingale mounted for the need to improve hospital hygiene.
2 What are the strengths of the polar graphs that Nightingale developed compared to other ways (e.g., tables) of displaying the same information?

Case study 8.2 A sustainability or environmental studies (science or geography) task

One tonne of carbon dioxide is the size of a balloon 12 metres across (Figure 8.8).

Figure 8.8 Balloon filled with one tonne of CO_2
Source: © Gary Braasch.

Questions
1. Would this balloon fit into your classroom? Explain your answer.
2. If it would not, try to explain where the balloon might fit.

A Melbourne-based family of four is planning to travel to the Gold Coast for a three-week holiday. The family is environmentally conscious. They are deciding whether to fly or drive. While there may be cost differences associated with the travel, they are more concerned about their carbon footprint in making the journey to and from the

Gold Coast. Below is the information they have regarding carbon dioxide emissions associated with flying and with driving:

- Flying distance is 1345 kilometres. One person's share of the emissions from the plane journey is 0.18 kilograms of carbon dioxide per kilometre flown.
- Driving distance is 1720 kilometres. The family owns a car which uses 7 litres of petrol for each 100 kilometres driven on the open road. For each litre of petrol, 2.50 kilograms of carbon dioxide are emitted.

Question

To minimise their carbon footprint, should the family fly or drive? To answer this question, you will need to work out how many 12-metre-diameter balloons (tonnes) of carbon dioxide will be emitted by the family if they fly and if they drive to the Gold Coast and back.

THE 21ST CENTURY NUMERACY MODEL AND ASSESSING NUMERACY

The 21st Century Numeracy Model can assist in the development of appropriate tasks that provide opportunities for teachers to assess how well their students' numeracy capabilities are being developed within the discipline they teach. In other chapters in this book, you have learned about the design of effective numeracy tasks. The focus in this chapter is on assessing numeracy capabilities.

Once a context is decided upon, there are three essential dimensions of the 21st Century Numeracy Model to consider when designing assessment tasks or activities: mathematical knowledge, tools and dispositions. A fourth dimension to consider is the inclusion of a critical orientation; as discussed elsewhere in this book, this dimension can prove challenging.

Assessment can be formal or informal. Informal assessment takes place when teachers observe children's behaviours, pay attention to what they say and record their impressions mentally or in note form. Formal assessment usually involves having data (most often in writing) to support conclusions drawn; testing is probably the most common form of formal assessment in classrooms.

Mathematical skills and tool use

There are fairly traditional means by which the mathematical skills exhibited by students can be assessed. Considerations include:

- Did the student understand what mathematical skills to draw on?
- Was the method used to find the answer to the mathematical dimension of the task appropriate?
- Was a correct solution achieved?
- Can the student explain how the answer was arrived at?

With respect to the use of tools, considerations include:

- Was an appropriate tool selected?
- Can the student explain why this tool was chosen?
- Was the tool used appropriately?
- Was the measure or outcome from the tool's use accurate or correct?

Dispositions

As in other aspects of educational practice, dispositions (attitudes and/or beliefs) are frequently regarded less seriously than cognition (or thinking). In many research studies, however, a correlation (positive relationship) has been found between achievement and a range of affective (attitudinal) factors, including enjoyment, interest, perceived usefulness of task and self-confidence (e.g., Chen et al. 2018; McLeod 1992; Thomson et al. 2003). Dispositions, including confidence, flexibility, initiative and risk are frequently overlooked with respect to learning outcomes in general. Dispositions are also notoriously difficult to assess in the classroom (e.g., Leder & Forgasz 2002). When observing what children are doing in the classroom or in written tests, it is relatively easy to determine if they can demonstrate (in writing, orally, or both) their understanding of concepts, as well as their competence with using various tools. So, how can dispositions be gauged?

Below are a few ideas for the informal and formal assessment of dispositions related to tasks in any subject domain that involves numeracy. Select at least one of these approaches and try it out with some students you are teaching or observing.

1. Observe students as they work on a numeracy task.
   - Do they appear engaged in the task?
   - Do they appear excited and enjoying what they are doing?
   - Are they trying different approaches (including different tools) to tackle the task, in their efforts to find answers (flexible)?
   - Are they seeking additional information and are they prepared to try something new or unknown (risk-taking)?

- Are they reticent and frequently seeking support or approval for their ideas (lacking in confidence)?
2 In a formal assessment of a task including a numeracy component, you could ask:
   a) Circle the words (one or more) which best describe how you felt as you were doing the task. Please add any other words to the list which describe how you felt.
   - pleased
   - clever
   - uninvolved
   - worried
   - unhappy
   - confused
   - enthusiastic
   - bored
   - great
   - excited
   - depressed
   - satisfied
   - stupid
   - interested
   - frustrated

   b) Circle the words (one or more) which best describe the task. Please add any other words to the list which describe the task.
   - fun
   - very easy
   - exciting
   - interesting
   - challenging
   - disappointing
   - boring
   - different
   - hard
   - same as always

For primary-aged students you could provide something like the image in Figure 8.9 and ask them to circle the face that most looks like how they felt as they worked on the task.

In a recent research study in which teachers learned about numeracy and how to seize opportunities to develop numeracy

Figure 8.9 Faces for assessing numeracy dispositions

activities across all discipline areas, we challenged teachers to gauge students' dispositions as they engaged with tasks. We interviewed the participating teachers on several occasions and asked questions about developing tasks and gauging students' dispositions towards them. Lily was one of the participating teachers. In the first two years of the study, she taught Year 9 students, and in the final year of the study, she was teaching Year 4 students. Extracts from the interview transcripts of some of Lily's comments about dispositions are reproduced below. The italicised words illustrate that Lily had thought about dispositions, that the students were exhibiting the dispositions encompassed in the 21st Century Numeracy Model, or both. The extracts reveal how teachers can relinquish control in the classroom and allow students to develop and demonstrate the dispositions needed to accomplish numeracy tasks. Also demonstrated in these extracts is that teachers need to be consciously aware of their students' dispositions, that careful observation is needed and that dispositions can be assessed in written tasks.

> **Interviewer** So the context is obvious—you've got them outside measuring things in the real world—and the mathematical knowledge is obvious, because it's about trigonometry. *Dispositions*: you've clearly thought about that, because you're

outside, the kids are engaged, and they're using mathematics to solve a life-related problem.

Lily And I also didn't give a lot of direction. Like when I presented the problems I just basically said, 'Use trigonometry to measure the height of a tree'. And *I wanted to keep it flexible* so that students actually could have a think about 'How do I do that?'

Interviewer Is that a new approach for you?

Lily That's pretty new. I possibly could've even taken it further and maybe got the students to design their own equipment to do it. That wasn't something I felt comfortable doing; I thought I just want to make sure that they've got the equipment and they know what they're doing. I did prompt in some of the ways that I was questioning: 'Well what do you need? What measurements would you need? You'd need to know the distance between . . .' Well, I didn't say that, but I left that as an open question. So I sort of was using questioning to get them to think about it, *but they ultimately came up with the ideas* . . . I saw them measure their angle and step away and have a conversation and then realise that they hadn't actually measured the distance, and then they'd have to say, 'Well, I've got to go and measure the angle again now because I don't know where I was standing'. *So they were able to be flexible and learned from their mistakes.*

Interviewer So that's you stepping back a bit. How did you feel about all that? Were you comfortable or did you have almost an overwhelming urge to rush in?

Lily Yes, I probably did have that urge, definitely. I suppose with my questioning, even, sometimes I think was I questioning too much and wanting them to get onto task and know what they're doing. Those boys ... I felt like they were just taking their time and not really getting into it, but then I didn't go up, even though I was feeling like they've done one and they've got to do at least four, but *they were the only ones who actually did an angle of depression* out in the yard. So I was kind of glad that I hadn't gone and told them what to do, because I probably would've told them to go and measure the goalposts or to do a building, but they actually came up with something that I wouldn't have actually suggested. *So that was a nice surprise for me, because I thought if I wasn't in there controlling them, they actually came up with something better than what I was going to suggest anyway. So that was really nice, actually.*

Interviewer A different experience.

Lily It was a different experience. I mean, I knew they were good kids and that they would do the right thing—I didn't feel like they were going to be messing about being silly—but it's hard to step back and relinquish control to a class and yet sometimes when you do it you realise that they actually will go and do what you want them to.

Interviewer Sometimes they get it right.

Lily I've got the assessment rubric and probably I couldn't think of what the proportion is of the mark, but maybe 5 per cent or 10 per cent is based around this reflection ... And also *one of my critical questions at the very end, my last post was 'Why do you think confidence, flexibility, initiative and risk are important in solving problems?'* So, just for them to sort of start to see ...

Interviewer To start to think about the *dispositions*?

Lily Yeah, and *why they need to sort of take risks.*

Interviewer And even I was saying to one of your students, 'What could be an example? *Let's think of an example where you could demonstrate confidence…*'

Lily *Yeah and often they don't see it*… I had one girl said, *'Oh, I don't think that I've demonstrated confidence. I'm not feeling…'* And I said, *'Well, what about when you … worked ahead* or when you …?' And then, 'Oh yeah, I did, didn't I? Oh.' *'Well did you research something without coming and asking me for the answer?'* 'Oh yes, I did.' *'Oh okay, you demonstrated initiative.'*

Interviewer And that's *risk-taking* as well.

Lily Yes, so I think just to highlight to them what they're demonstrating.

Lily They also were asked to demonstrate how they showed confidence, flexibility and initiative and took risks. They also had to give me examples of the tools that they used and the concepts, skills and problem-solving that they did. They had to create like a concept map showing me all of those, where they actually showed those aspects of the model within their project. So they had to say, 'When I was doing this part of the assignment I had a lot of trouble and I showed initiative by actually speaking to my mum or my auntie who is an accountant; she was able to help me' or 'I showed initiative by seeing the teacher after class.'

Interviewer Or persisted.

Lily 'I took a risk maybe because …' I'm just trying to think of some of the responses. 'I didn't think I knew how to do it, but I had a go and I was able to nut the answer out, and

I showed confidence because I could actually, as the tasks went, I actually was more confident.' So it gave the kids that opportunity to reflect on their own dispositions and allowed them to discuss why they were important as well, and that then became part of the assessment as well.

Interviewer What were you attempting to find out about student learning when you implemented that assessment task and did it in that way?

Lily I suppose their ability to identify and have that opportunity to discuss with them, as well. Like, I think a lot of them didn't really know. Like, I could see that they'd been flexible but they could not identify that... So I think it gave them that opportunity to reflect on their own practice and to identify 'Oh, that's a behaviour that is important in numeracy'. That is something that I think if I had have done that again and again and again the kids would have started to realise, 'Oh, I'm showing flexibility right now. Oh okay, I'm having trouble. Yesterday, I showed flexibility and that really helped. I'm going to try and be flexible now.' So just sort of reading or building those behaviours and placing value upon them.

Interviewer Were there other things that you have attempted to assess even beyond the dimensions of the numeracy model—have you gone further than that?

Lily Well, in the primary classroom, obviously doing lots more observational stuff as well, like looking at confidence, and yes, I think just sort of more taking observations...

Interviewer As you were saying, that student who was already walking back from a task, yes, talking about the task and really, already thinking that kind of stuff.

Lily Yes, that's right, and he's showing initiative already, because he's thinking already about it. And that's something he'll probably consider in the next couple of days. When we come back to it he'd probably be one of my ... I can go straight to him and open up that conversation as well.

> ### Review and reflect 8.7
>
> 1 Implement with your students one of the numeracy tasks found in this book.
> 2 Gather evidence of the students' dispositions towards numeracy as they engage with the task.
> 3 What can you conclude about the students' dispositions?

CONCLUSION

Numeracy assessment can have different meanings and implications. At the national and international levels, the results of large-scale testing programs are used to evaluate and compare the performances of schools and education systems. However, it is important to recognise that in these testing regimes, different definitions of numeracy (or mathematical literacy) are used. In Australia's annual NAPLAN test, it is school students' mathematical knowledge (as outlined in the Australian Curriculum—Mathematics), and not their numeracy capabilities (as defined in the Australian Curriculum) that is being assessed.

Assessment of numeracy is typically associated with formal testing programs that are devised, administered and interpreted by authorities outside the school setting, and so it would appear that teachers have little role to play in assessing their own students' numeracy development. However, teachers can contribute their expertise in two main ways. First, they can work with colleagues in their schools to interpret NAPLAN results for individual students and at the pertinent grade-levels, and use this information to modify their teaching approaches so as to address areas of weakness. Second, teachers of all school subjects can use a wide range of informal assessment techniques that allow students to demonstrate their numeracy capabilities—not only in terms of mathematical knowledge relevant to learning the subject, but also through the appropriate use of tools and development of positive dispositions.

RECOMMENDED READING

Australian Council for Educational Research, 2017, *Literacy and Numeracy Test for Initial Teacher Education Students: Sample questions*, <https://teacheredtest.acer.edu.au/files/Literacy_and_Numeracy_Test_for_Initial_Teacher_Education_students_-_Sample_Questions.pdf>, retrieved 11 March 2018

National Assessment Program, 2016, *Numeracy*, <www.nap.edu.au/naplan/numeracy>, retrieved 11 May 2018

Organisation for Economic Co-operation and Development, 2018, *Survey of Adult Skills. First results; Australia, Country note*, <www.oecd.org/skills/piaac/Country%20note%20-%20Australia_final.pdf>, retrieved 6 July 2018

Thomson, S, De Bortoli, L. & Underwood, C., 2017, *PISA 2015: Reporting Australia's results*, <http://research.acer.edu.au/cgi/viewcontent.cgi?article=1023&context=ozpisa>, retrieved 11 March 2018

9

Challenges in enhancing numeracy

There are many challenges in planning for and promoting numeracy learning across the school curriculum (Hughes-Hallett 2001). For example, in Chapters 4 and 5, we saw how teachers need to develop the capacity to recognise where numeracy demands and opportunities exist in the subjects they teach. This chapter provides an opportunity for you to reflect on how well prepared you are to promote numeracy learning and to begin to frame how you will continue to develop your capacity to embed numeracy into the subjects you are teaching. To provide a context for your reflection and future planning, we evaluate the experiences of Australian teachers in embedding numeracy across the curriculum and of pre-service teachers who have undertaken a compulsory course on numeracy across the curriculum as part of their initial teacher education program.

In the first section of the chapter, we consider the ways in which teachers' personal conceptions of numeracy can change over time. Next, we explore some of the challenges facing teachers who wish to embed numeracy into the subjects they teach. Then, we introduce a framework for identity as an embedder-of-numeracy (Bennison 2017) to assist you to reflect on how well prepared you are to embed numeracy into the subjects you will teach. Finally, we look at some perspectives on numeracy of students and pre-service teachers.

PERSONAL CONCEPTIONS OF NUMERACY

Teachers must be able to model the kind of numeracy they want their students to develop if they are to effectively promote numeracy learning through the subjects they teach. To do this, they need to have a rich personal conception of numeracy that encompasses all of the dimensions of numeracy seen in the 21st Century Numeracy Model (Figure 3.1). In Review and reflect 1.1, we asked you to write down your responses to four prompts to capture your ideas about numeracy. We also saw some of the responses to the same task made by the first group of teachers we worked with (see Goos, Geiger & Dole 2011), who completed the task at the beginning and end of their project. Their responses at the end of the project demonstrated a more sophisticated understanding of numeracy than those at the beginning. They understood numeracy as involving the confident application of mathematical knowledge, skills and problem-solving strategies across a range of everyday contexts. However, references to the use of tools (e.g., 'can use a variety of tools') and the need to apply a critical orientation (e.g., 'knows when data have been manipulated to present bias') were made much less frequently than comments about other dimensions of the numeracy model. Our classroom observations revealed increasing use of tools, especially digital technologies such as spreadsheets. However, classroom observations and interviews confirmed that teachers found critical orientation the most difficult aspect of the numeracy model to implement in classroom activities.

The challenge of incorporating a critical orientation when designing tasks for implementation in classrooms is seen in the responses to a version of the numeracy trajectory task of another group of teachers

> ### Review and reflect 9.1
>
> 1 Has your personal conception of numeracy changed as you have worked through this book? Write down once more your responses to the prompts in Review and reflect 1.1 to capture your current ideas about numeracy.
> 2 Compare your responses with those you first gave. You can analyse your responses by matching them to the dimensions of the 21st Century Numeracy Model (Figure 3.1). For example, 'Numeracy involves using mathematics to be successful in everyday life' can be matched to the mathematical knowledge and contexts dimensions of the numeracy model (for more examples, see Goos, Geiger & Dole 2011).
> 3 Compare your responses with those of a partner and summarise any similarities and differences you identify.

we have worked with (Geiger, Forgasz et al. 2015). As we explained in Chapter 7, the task was designed to capture the teachers' foci on the different dimensions of the numeracy model. We asked teachers to annotate a copy of the numeracy model to map their trajectory through the model when designing a numeracy task. We also asked them to identify their initial focus and then the order in which they incorporated other dimensions of the numeracy model. Of the ten teachers who completed the task, seven identified mathematical knowledge as the dimension they first considered when thinking about how to design a numeracy task. Only one teacher first thought about a critical orientation. One teacher did not identify the critical orientation as a priority, and four out of the ten teachers identified the critical orientation as the last dimension to be incorporated.

> **Review and reflect 9.2**
>
> 1 Think about a numeracy task you have designed. Using the copy of the 21st Century Numeracy Model in Figure 3.1, annotate the dimension that was your initial focus and then the order in which you incorporated the other dimensions of the model.
> 2 Compare your numeracy trajectory sequence for the development of this task with the sequences provided by the teachers in our project (see Geiger, Forgasz et al. 2015, p. 616) and with the sequence used by a partner.
> 3 Discuss possible reasons for the use of different sequences and consider whether there is a 'correct' sequence. Are all dimensions of numeracy relevant to every task?

In our work with teachers, we have found that the numeracy model can support the framing of activities so that greater attention is paid to all dimensions of numeracy, especially the critical dimension, but that time and experience are needed for teachers to become effective designers of numeracy tasks. As you move from being a pre-service teacher to a beginning teacher, much of your learning in all aspects of teaching is likely to occur when you reflect on lessons. Your capacity to design effective numeracy tasks will improve over time, as long as you look at ways to enrich the activities you use. One early career teacher we have worked with, when talking about how to incorporate a numeracy focus into activities, commented, 'It's more reflective stuff, like it's more that I do the activity and later on think, "Oh, you know what would have been really great for that activity is to actually do this and this, and maybe this."'

CHALLENGES TO ACROSS-THE-CURRICULUM NUMERACY

Three of the challenges facing teachers who wish to take an across-the-curriculum approach to numeracy are the disciplinary boundaries that exist in the Australian Curriculum (ACARA 2018a), the increasing emphasis on science, technology, engineering and mathematics (STEM; see ACARA 2018f) and the lack of resources available to support teachers' understanding and enactment of numeracy across the curriculum.

Australian Curriculum

There are eight learning areas in the Australian Curriculum (ACARA 2018a): English, mathematics, science, humanities and social sciences, the arts, technologies, HPE and languages. These learning areas provide disciplinary boundaries that can seem at odds with the cross-curricular nature of the seven general capabilities that are to be addressed through all learning areas: literacy, numeracy, information and communication technology, critical and creative thinking, personal and social capability, ethical understanding and intercultural understanding. The existence of disciplinary boundaries seems to be particularly problematic for numeracy, where the elements of the ACARA's numeracy learning continuum are expressed in terms of mathematical content (e.g., 'Recognising and using patterns and relationships'). At face value, these appear to have little relevance outside the mathematics classroom. As a result, teachers of non-mathematics subjects may feel that the intent of including numeracy across the curriculum is to make every teacher a teacher of mathematics, and mathematics teachers may feel that there is nothing additional for them to consider. We now reflect on both of these perceptions.

For non-mathematics teachers, the dilemma is to manage the demands of teaching a subject and to embed numeracy into that subject without losing focus on the subject they are teaching. There have been many examples presented in this book that illustrate how attending to numeracy in subjects across the curriculum can actually enhance subject learning (see Chapters 4 and 5). Being able to recognise these demands and opportunities will help to break down disciplinary barriers that prevent students from making connections between their learning in different subjects and also to enhance students' disciplinary understandings.

To be numerate, a person needs to be able to use mathematics in a range of contexts. Thus, the challenge for mathematics teachers is to pay particular attention to how mathematics is used beyond the mathematics classroom. This can be achieved by providing problems for which the solution depends on the context and asking students to justify their solutions and the choice of the mathematical skills they use. The contexts chosen by the teacher can come from various learning domains or from situations of relevance to students.

STEM

STEM disciplines have a pivotal role in making Australia globally competitive (Office of the Chief Scientist 2014). However, the performance of Australian students in mathematical literacy declined between the 2003 and 2015 PISA cycles (Thomson et al. 2017) and the number of secondary school students studying calculus-based mathematics subjects has also declined over a similar timeframe (Barrington & Evans 2016). This suggests that many students are leaving the Australian schooling system ill equipped to pursue tertiary study in the STEM disciplines. These and related concerns have led to a national STEM school education strategy (Education Council 2015).

The increased focus on STEM education in the Australian Curriculum (ACARA 2018f), with an integrated approach to the teaching and learning of mathematics, science and technologies, can be seen as a new agenda that competes with the goal of improving students' numeracy capabilities more broadly. We consider the two agendas as complementary, because improving students' numeracy capabilities through an across-the-curriculum approach highlights the relevance of mathematical knowledge in understanding other disciplines as well as the world beyond the mathematics classroom (i.e., real-world contexts). As a result, students may be more likely to consider studying advanced- and intermediate-level mathematics subjects (that include the study of calculus) in their senior secondary years, thereby preparing themselves for future careers in the STEM disciplines. One of the consequences of the increasing importance of STEM is the need for citizens to be able to make informed decisions about the technological advances taking place, based on data presented to them. In order to do this effectively, all students leaving school need an appropriate level of numeracy, especially the capacity to apply a critical orientation.

Resources for numeracy teaching and learning

In Chapter 6, we discussed principles for the design and implementation of numeracy tasks, demonstrating that teachers can create such tasks as well as selecting or adapting tasks from available resources. Nevertheless, teachers who are just beginning to engage with the idea of numeracy across the curriculum often benefit from working with existing tasks that have been designed for this purpose. As discussed in detail in Chapter 7, to find out what materials were available to teachers, we conducted an audit of resources that support teachers to understand and enact numeracy across the curriculum (Goos, Geiger, Bennison et al. 2015). Through our audit, we found very few resources that addressed the need for teachers to recognise and take advantage of the numeracy demands and opportunities within the subjects they teach.

In response to our findings from the resource audit and from a gap analysis informed by interviews with relevant staff from stakeholder groups—including employing authorities and teacher professional associations—we developed six videos (University of Queensland 2018). In Chapters 5 and 7 of this book, you were introduced to these videos from the perspectives of numeracy task design and approaches to developing a whole-school approach to numeracy. Four of the videos illustrate how teachers embed numeracy in the subjects they teach, one shows teachers discussing how they established a numeracy committee within their school, and one is in the form of a voiceover PowerPoint, in which the 21st Century Numeracy Model is explained. Discussion questions designed to engage viewers with an across-the-curriculum approach to numeracy accompany each video.

EMBEDDERS-OF-NUMERACY

Becoming a teacher involves more than acquiring the necessary knowledge and skills to teach particular subjects. Those who are new to the profession must also develop an understanding of what it means to be a teacher (Sachs 2005). This process involves developing a *teacher identity*, which, when discussing teacher education programs and the types of teacher identities they promote, Sachs described in the following manner: 'Teacher professional identity then stands at the core of the teaching profession. It provides a framework for teachers to construct their own ideas of "how to be", "how to act", and "how to understand" their work and their place in society. Importantly, teacher identity is not something that is fixed nor is it imposed; rather it is negotiated through experience and the sense that is made of that experience' (p. 15). This portrayal of teacher identity highlights that identity development is ongoing, that it involves both the person and context, and that individuals

> **Review and reflect 9.3**
>
> Choose one of the three challenges of an across-the-curriculum approach to numeracy listed below. Provide a written response to the prompt following the challenge to develop your own position in relation to it. Defend your position in discussion with a colleague.
>
> - Disciplinary boundaries in the Australian Curriculum: It is impossible to cover all the content in the separate learning areas in the Australian Curriculum at the same time as embedding numeracy across the whole curriculum. Priority should be given to learning area content, and numeracy should just be taught in mathematics lessons.
> - The rise of the STEM agenda: The STEM agenda has overtaken numeracy. Schools cannot focus on both, so STEM should be the priority area, because it prepares students for further study and careers.
> - The lack of available numeracy resources: There are not enough quality resources available for helping teachers to embed numeracy across the curriculum. Teachers cannot be expected to develop their own numeracy resources; instead, they should concentrate on teaching the learning areas in the Australian Curriculum.

can exercise agency. Sachs did not mention that individuals have multiple identities which stem from the practices and contexts of the communities in which they participate (Wenger 1998), possibly because her focus was on teachers' overall, or core, identity (Gee 2001). We maintain that one of the multiple identities of teachers is their identity in the context of promoting numeracy learning through the subjects they teach.

Teacher identity is seen by many researchers as providing useful insights into the learning and practices of teachers (e.g., Enyedy et al. 2005; Goodnough 2011; Graven 2004). However, many factors contribute to a teacher's identity, and, for this reason, defining teacher identity in a way that makes this construct operational is one of the challenges of using identity in empirical research (Sfard & Prusak 2005). By invoking the situated nature of identity (Wenger 1998), it is possible to identify the knowledge and affective attributes, social interactions and environmental factors that contribute to shaping how teachers address numeracy in the subjects they teach. The framework for identity as an embedder-of-numeracy described by Bennison (2017) is organised by five *domains of influence*—life history, knowledge, affective, social and context—and includes factors that are likely to influence how teachers promote numeracy learning through the subjects they teach (see Figure 9.1). We will now look more closely at the elements that contribute to each domain.

Life history domain

Past experiences contribute to teachers' identity development (e.g., Philipp 2007). Consequently, many factors that influence how you promote numeracy learning are likely to have been shaped by your past experiences.

- What is the nature (positive or negative) of your past experiences of mathematics, and what opportunities have you had

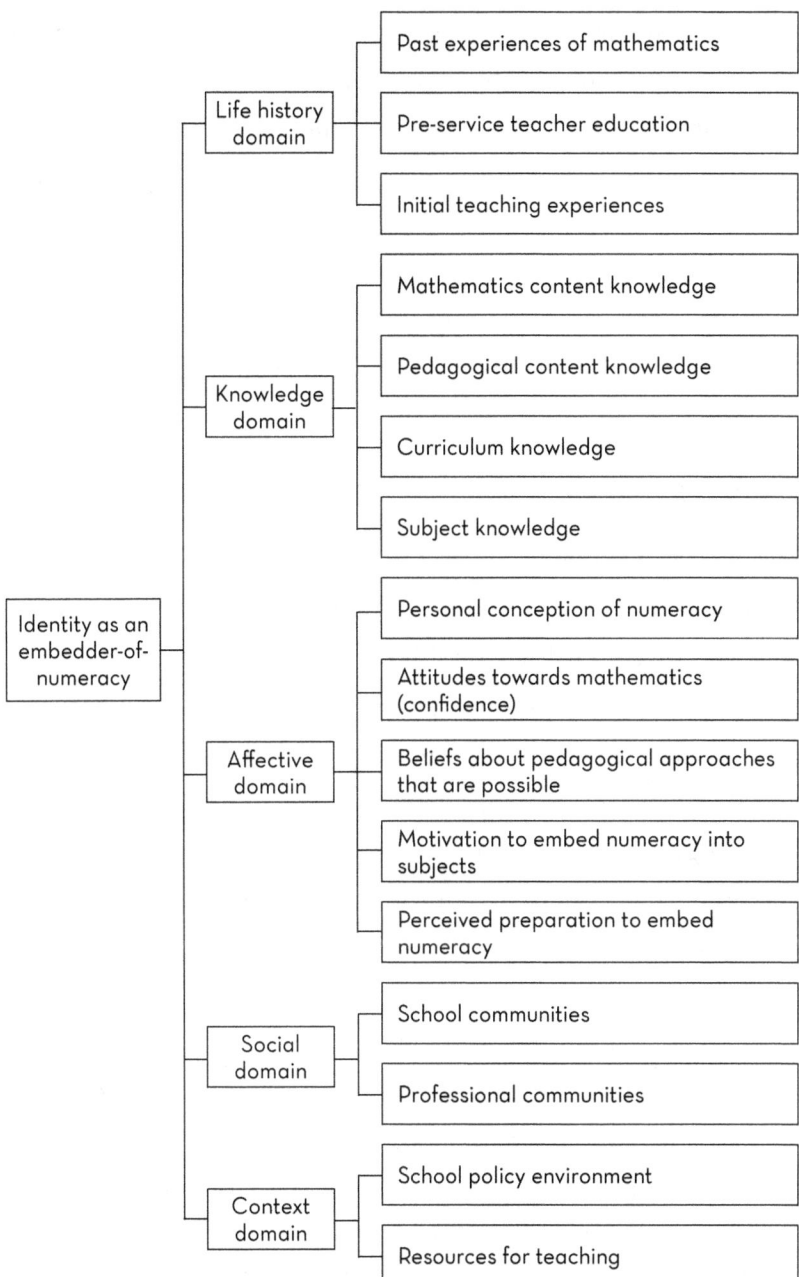

Figure 9.1 A framework for identity as an embedder-of-numeracy
Source: Bennison 2017.

(formal and informal) to develop competency with the inherent mathematics in the subjects you teach?
- What opportunities did you have during pre-service teacher education to learn about how numeracy can support subject learning and develop pedagogical content knowledge for numeracy?
- In your professional practice, what opportunities have you had to engage with an across-the-curriculum approach to numeracy?

Knowledge domain

Your professional knowledge is an important part of your identity as a teacher (e.g., Van Zoest & Bohl 2005). Several types of knowledge are needed for teaching (Shulman 1987), but only four types are included in the knowledge domain (mathematical content, pedagogical content, curriculum and subject knowledge). Subject knowledge (encompassing content, pedagogical and curriculum knowledge) is typically developed through pre-service teacher education courses. However, it should be noted that there is some overlap in the four types.

- Have you developed the appropriate level of mathematical content knowledge to deal with the mathematics inherent in the subjects you teach?
- Have you developed the pedagogical content knowledge to be able to design effective numeracy tasks?
- Have you developed the curriculum knowledge to identify numeracy learning opportunities and make connections between numeracy and subject learning?

Affective domain

Affective attributes cover a broad spectrum (e.g., Philipp 2007), so the factors included in this domain could also be considered as part of other domains. We have already considered beliefs about what numeracy encompasses earlier in this chapter, but there are other affective attributes that influence your capacity and desire to promote numeracy learning.

- How confident are you with the inherent mathematics in the subjects you teach?
- What connections can you see between numeracy and subject learning that might motivate you to address numeracy in the subjects you teach?
- How well prepared do you feel to address numeracy in the subjects you teach?

Social domain

Identity development involves participation in communities (Wenger 1998). As a teacher you participate in many communities, both within and outside the school. Interactions you have with others in these communities have the potential to influence the way you promote numeracy learning. For example, interactions with colleagues and school administrators may involve discussions about the meaning of numeracy and who is responsible for numeracy learning. These discussions may support an across-the-curriculum approach to numeracy, as has been described elsewhere in this book. Alternatively, you may need to be able to articulate the benefits of such an approach to colleagues and administrators in your school community. Participation in professional communities has the potential to further develop your capacity to promote numeracy learning through the subjects you teach.

- What interactions have you had with other teachers and administrators in your school about numeracy across the curriculum? What have you learned about their perceptions of numeracy and who is responsible for developing students' numeracy?
- What opportunities exist in teacher professional associations for you to learn more about embedding numeracy into the subject you teach?

Context domain

Practice and identity are related (Wenger 1998), so affordances for and constraints on practice within your professional context can influence the ways in which you promote numeracy learning. The school policy environment includes curriculum initiatives and accountability measures related to numeracy and so can offer both affordances and constraints. For example, a whole-school approach to numeracy across the curriculum can help you to effectively address numeracy in the subjects you teach. On the other hand, undue emphasis on preparation for national testing could potentially limit your views on what numeracy is and what it means to be numerate. Being able to effectively use representational, physical and digital tools is an integral part of being numerate. Consequently, it is advantageous to have access to appropriate resources for teaching if you are to promote numeracy learning through the subjects you teach.

- What affordances for and constraints on practice in relation to embedding numeracy across the curriculum exist in your professional context?
- Do you have access to adequate resources for teaching in order to promote numeracy learning in the subjects you teach?

> **Review and reflect 9.4**
>
> A beginning teacher's identity as an embedder-of-numeracy will have mainly been shaped by elements within the life history, knowledge and affective domains. As one's career progresses, elements within the social and context domains are likely to become more influential in identity development.
>
> 1. Write a paragraph responding to the questions posed in the sections on life history, knowledge and affective domains. How do you see your current identity as an embedder-of-numeracy?
> 2. Add to the paragraph any information relevant to your experience in the social and context domains.
> 3. Discuss your responses to the questions in this section with a partner.

STUDENTS' PERSPECTIVES ON ACROSS-THE-CURRICULUM NUMERACY

During school visits that have been part of our work with teachers on numeracy, we have observed lessons and interviewed teachers and students. We have asked students to reflect on the numeracy lessons they have recently experienced and to express their feelings about mathematics and using mathematics in various activities (e.g., Geiger, Goos & Dole 2014). One group of students interviewed after a Year 8 (12–13 years of age) HPE lesson could identify 'mathematics' in other subject areas (contexts). The teacher in this instance was working in a middle school that required her to teach across the curriculum, perhaps helping students to make connections as a result of her participation in the study.

> **Student 1** That's what our teacher, Mrs C—, is trying to do. Trying to get more maths stuff in other subjects.
> **Researcher** Do you think that it is working? Can you see how maths is being integrated into other subjects?
> **Students (together)** Yeah.
> **Student 2** And I think HPE is the best lesson to do it in as well, because all of your sports and stuff, you use numbers for your scores. And if you can learn it in different areas as well, it helps.
> **Researcher** What about other subject areas? Can you tell us about some examples where Mrs C— has put more maths there?
> **Student 1** In science earlier this year, we done a test to see how much one plain peanut, how much exercise you have to do to work off that much.

With prompting, students were able to make connections between mathematics and other discipline areas. Examples are provided below:

English

Student 3   Yeah, remember we were doing that thing about how many pence in a dollar. Remember reading the *Red Dog* book?

Student 1   Yeah. When we listened to the recording of *A Fortunate Life* (a novel), it was talking about the old currencies, and we were trying to do some maths on the amounts of pence and pounds.

Home economics

Student 2   We're doing a bit of sewing. Tomorrow we're making boxer shorts.

Student 1   Yeah . . . lots and lots of measuring.

Society and the environment

Student 1   And we recently did a section on koalas and the different types of trees they eat. And we did a percentage table of the different types of trees eaten by a koala. And we did a pie graph for that.

In the observed HPE lesson, these students had used pedometers (a physical tool) to collect data and an Excel spreadsheet (a digital tool) to record, represent and analyse the data they had collected.

Researcher   But did you notice, when you were looking at that on different days of the week, each of you were walking different numbers of steps?

Student 2   Yeah. Sunday was the smallest.

Researcher   I noticed!

Student 1 I was going to say that Thursday and Saturday probably would have been the biggest too, 'cause that's when we play sport.

Student 2 We did some graphs on the computer too, showing two days, and I did Saturday and Sunday on a line graph … and there was a major difference! Saturday was like this [gesturing to show a large number of steps] and Sunday was like this [gesturing to show a small number of steps].

The above exchange highlights students' use of a critical orientation to draw conclusions about their own and other class members' levels of physical activity. There was also evidence in this interview of students' willingness to take the lead in assisting other members of the class (dispositions): 'Well, some of us are a bit better with it [using Excel]—like we understand it a lot quicker, so just gave them some help.'

ACROSS-THE-CURRICULUM NUMERACY IN INITIAL TEACHER EDUCATION

In Chapter 8 we discussed implications for graduate teachers of the Australian Professional Standards for Teachers (AITSL 2017) and the mandated literacy and numeracy test (LANTITE; ACER 2017a).

The Australian Curriculum expectation that teachers develop students' numeracy capabilities across the curriculum and the Australian Institute for Teaching and School Leadership's standards for graduate teachers' personal numeracy skills underpinned the development of the unit 'Numeracy for learners and teachers' at one Australian university. The 21st Century Numeracy Model framed the content and approach to the teaching of the unit. 'Numeracy for learners and teachers' was first offered in 2015 as a compulsory unit in four streams (primary, early years and primary, primary and secondary, and secondary) of a two-year master of teaching program. The course was not focused on teaching numeracy skills to assist students to pass the numeracy part of LANTITE. Rather, students were expected to have the necessary background mathematical skills to deal with and capitalise on the numeracy opportunities within the subjects they would teach in future and to meet the numeracy demands of their future workplaces—schools. Support was provided via online 'self-help kiosks' for those who wanted to sharpen their background mathematical skills.

To explore the impact of participation in the 'Numeracy for learners and teachers' course on meeting Australian Institute for Teaching and School Leadership standards as well as the expectations of the Australian Curriculum, at the beginning and end of the unit, we evaluated students' understanding of the concept of numeracy as well as their confidence in embracing numeracy opportunities in the subjects they would teach. We gathered data from students enrolled in the unit in the three years from 2015 to 2017 (for a preliminary report, see Forgasz & Hall 2016). As well as basic biographical details, both the pre- and post-unit surveys asked the following questions:

Q1 Do you believe there are differences between mathematics and numeracy?

Q2 Are there mathematical demands on teachers in schools apart from what is taught to students?

In every year surveyed, a much higher proportion of students answered 'yes' in response to both questions in the post-unit survey than in the pre-unit survey. Experiencing 'Numeracy for learners and teachers' had changed students' views. The results of the two surveys are shown in Table 9.1.

Figure 9.2 shows the 2015–17 participants' reported pre- and post-unit levels of confidence in 'incorporating numeracy into the teaching of [their] subject area(s)'. There was a clear impact on this confidence by the time of completion of the unit. Before beginning the unit, many students reported being less than 'somewhat confident' (62 per cent in 2015, 54 per cent in 2016 and 36 per cent in 2017). After completing the unit, the vast majority of students reported being 'somewhat confident' or 'very confident' (100 per cent in 2015, 92 per cent in 2016 and 94 per cent in 2017).

Table 9.1 Results of 'Numeracy for learners and teachers' pre- and (post-) unit surveys

Participants	Years		
	2015 (n = 300, S 67%)	2016 (n = 140, P 67%)	2017 (n = 450, S 70%)
Number of participants	48 (21)	33 (13)	56 (17)
Female (%)	81 (74)	90 (81)	76 (71)
Aged 25-34 (%)	77 (74)	80 (86)	72 (79)
Prevalent course stream (%)	S 74 (S 80)	P 79 (P 90)	S 69 (S 61)
Did not study university mathematics (%)	66 (63)	78 (70)	70 (67)
Answered 'yes' to Q1 (%)	76 (95)	90 (92)	76 (93)
Answered 'yes' to Q2 (%)	64 (90)	75 (85)	59 (94)

Note: S secondary course stream, P primary course stream.

CHALLENGES IN ENHANCING NUMERACY

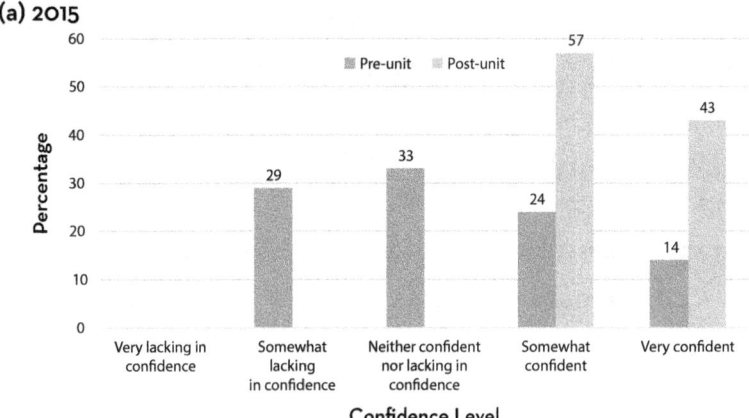

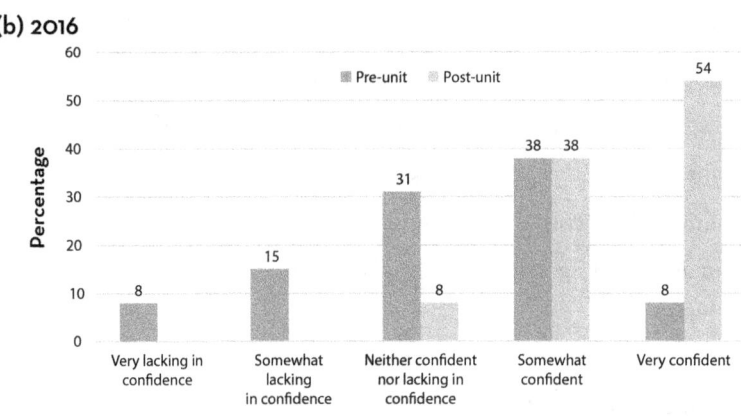

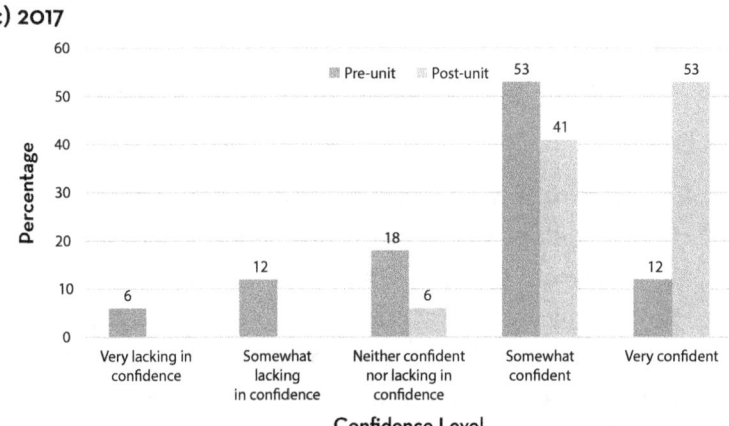

Figure 9.2 'Numeracy for learners and teachers' survey participants' pre- and post-unit confidence levels, 2015–17

Source: Graph for 2015 results adapted from Forgasz & Hall 2016.

Students were also asked to explain the confidence levels they reported. Examples of what students wrote included:

> I have a clearer understanding of what numeracy entails, have been provided examples with how it would work in my method curriculum areas, and feel confident that I have adequate mathematical reasoning and numeracy skills to be able to handle this in my teaching.

> I feel empowered and reassured that my 'average' knowledge of mathematics is enough to address it across the board of Primary school subjects. Through topic-based teaching I hope to engage in a variety of angles of approach, thus covering a number of curriculum demands at once. The demands of the mathematics curriculum can definitely be incorporated alongside those of other subjects such as humanities, literacy, PE and Health, science and so on.

> **Review and reflect 9.5**
>
> In the activity in Review and reflect 1.9, you evaluated your preparedness to address numeracy using the self-assessment survey we developed in our research with teachers.
>
> 1. Revisit and update your annotations to record your current understanding and confidence in embedding numeracy into the subjects you teach.
> 2. Identify any statements that you do not understand or feel confident in demonstrating.
> 3. Compare your findings with those of a partner and discuss how you could further develop your understanding and confidence in embedding numeracy into the subjects you teach.

CONCLUSION

There is no doubt that embedding numeracy across the school curriculum is a challenging enterprise. Although education researchers and policy-makers have offered substantial support for this goal for many years, teachers have lacked clear and practical guidance on how to recognise and exploit the numeracy demands and opportunities in the subjects they teach. At the institutional level, there have been few examples of how to implement a whole-school approach to numeracy across the curriculum. In this book, we hope that teachers will have found both the motivation and the tools to take up the challenge and support their students' numeracy development.

RECOMMENDED READING

Bennison, A., 2017, 'Re-examining a framework for teacher identity as an embedder-of-numeracy', in A. Downton, S. Livy & J. Hall (eds), *40 Years On: We are still learning*, Proceedings of the 40th annual conference of the Mathematics Education Group of Australasia, pp. 101–8, Melbourne: MERGA

Education Council, 2015, *National STEM School Education Strategy, 2016–2026*, <www.educationcouncil.edu.au/site/DefaultSite/filesystem/documents/National%20STEM%20School%20Education%20Strategy.pdf>, retrieved 11 March 2018

Forgasz, H. & Hall, J., 2016, 'Numeracy for learners and teachers: Evaluation of an MTeach coursework unit at Monash University', in B. White, M. Chinnappan & S. Trenholm, S. (eds), *Opening Up Mathematics Education Research*, Proceedings of the 39th annual conference of the Mathematics Education Research Group of Australasia, pp. 228–35, Adelaide: MERGA

Goos, M., Geiger, V. & Dole, S., 2014, 'Transforming professional practice in numeracy teaching', in Y. Li, E. Silver & S. Li (eds), *Transforming Mathematics Instruction: Multiple approaches and practices*, pp. 81–102, New York: Springer

Wenger, E., 1998, *Communities of Practice: Learning, meaning and identity*, Cambridge, UK: Cambridge University Press

ACKNOWLEDGEMENTS

Our goal in writing this book was to promote a rich interpretation of numeracy that connects the mathematics learned at school with out-of-school situations that additionally require problem-solving, critical judgement and making sense of the non-mathematical context. This approach necessarily positions numeracy as an across-the-curriculum commitment that extends beyond the mathematics classroom. The book is the culmination of a research and development program that builds on 17 years of productive and systematic engagement with teachers, teacher educators, policy-makers, school systems and the Australian and international research community. We thank all the people who have shared their numeracy education experiences and insights with us during this time.

We are grateful to the following for permission to reproduce material:

- Australian Curriculum, Assessment and Reporting Authority (Figures 5.1–5.3, 8.1). Licensed under Creative Commons
- Australian Government Department of Education and Training (Table 2.1, Figure 8.4)
- Lisa DeBortoli, Australian Council for Educational Research (Figure 8.2)
- Department for Education, South Australia (Department of Education and Children's Services, 2005, *South Australian Curriculum, Standards and Accountability Framework: Middle years band*, <www.sacsa.sa.edu.au/ATT/%7B85CFF734-68DE-

4F6D-A626-4EA1EDEC69C2%7D/SACSA_6_MYB.pdf>, retrieved 5 May 2017). Licensed under Creative Commons. 'This material is out of date since the introduction of the Australian Curriculum in 2012. The Government of South Australia, its agents, instrumentalities, officers and employees make no representations about the accuracy of the information and accepts no liability however arising from any loss resulting from the use of this information'

- Guardian News and Media Ltd
- Massachusetts Medical Society (Figure 3.2)
- Mathematics Education Research Group of Australasia (Figures 3.1, 7.2, 9.1, Table 3.1)
- Organisation for Economic Co-operation and Development (Figure 8.3)
- Joan Rothlein, Cedar Braasch and Gary Braasch Photography (Case study 8.2)
- Rick Roth for the poem 'So Fast' (Chapter 6)
- SpringerNature (Table 5.1)
- *The Sunday Mail*
- Steve Thornton
- Worldmapper.org (Figures 1.3 and 1.4)

REFERENCES

AAMT (Australian Association of Mathematics Teachers), 2006, *Standards for Excellence in Teaching Mathematics in Australian Schools*, <www.aamt.edu.au/standards>, retrieved 3 March 2018

ACARA (Australian Curriculum, Assessment and Reporting Authority), 2016, *NAPLAN Results*, <http://reports.acara.edu.au/>, retrieved 6 July 2018

ACARA (Australian Curriculum, Assessment and Reporting Authority), 2017, *National Literacy and Numeracy Program: NAP*, <www.nap.edu.au/home>, retrieved 28 December 2017

ACARA (Australian Curriculum, Assessment and Reporting Authority), 2018a, *Australian Curriculum*, <www.australiancurriculum.edu.au/>, retrieved 11 March 2018

ACARA (Australian Curriculum, Assessment and Reporting Authority), 2018b, *Australian Curriculum: English*, <www.australiancurriculum.edu.au/>, retrieved 9 August 2018

ACARA (Australian Curriculum, Assessment and Reporting Authority), 2018c, *Australian Curriculum: Health and physical education*, <www.australiancurriculum.edu.au/>, retrieved 9 August 2018

ACARA (Australian Curriculum, Assessment and Reporting Authority), 2018d, *Australian Curriculum: Humanities and social sciences (History)*, <www.australiancurriculum.edu.au/>, retrieved 9 August 2018

ACARA (Australian Curriculum, Assessment and Reporting Authority), 2018e, *Numeracy*, <www.australiancurriculum.edu.au/f-10-curriculum/general-capabilities/numeracy>, retrieved 11 March 2018

ACARA (Australian Curriculum, Assessment and Reporting Authority), 2018f, *STEM Report*, <www.australiancurriculum.edu.au/resources/stem/stem-report/>, retrieved 11 March 2018

ACER (Australian Council for Educational Research), 2017a, *Literacy and Numeracy Test for Initial Teacher Education Students*, <https://teacheredtest.acer.edu.au/>, retrieved 11 March 2018

ACER (Australian Council for Educational Research), 2017b, *Prepare*, <https://teacheredtest.acer.edu.au/prepare>, retrieved 11 March 2018

ACER (Australian Council for Educational Research), 2018a, *Literacy and Numeracy Test for Initial Teacher Education Students: Sample questions*, <https://teacheredtest.acer.edu.au/files/

Literacy_and_Numeracy_Test_for_Initial_Teacher_Education_students_-_Sample_Questions.pdf>, retrieved 11 March 2018
ACER (Australian Council for Educational Research), 2018b, *OECD Programme of International Student Assessment (PISA Australia)*, <http://research.acer.edu.au/ozpisa/>, retrieved 9 July 2018
ACER (Australian Council for Educational Research), 2018c, *Trends in International Mathematics and Science Study (TIMSS)*, <https://www.acer.org/timss>, retrieved 9 July 2018
AITSL (Australian Institute for Teaching and School Leadership), 2015, *Accreditation of Initial Teacher Education Programs in Australia: Standards and procedures*, <www.aitsl.edu.au/docs/default-source/initial-teacher-education-resources/accreditation-of-ite-programs-in-australia.pdf>, retrieved 9 August 2018
AITSL (Australian Institute for Teaching and School Leadership), 2017, *Australian Professional Standards for Teachers*, <www.aitsl.edu.au/teach/standards>, retrieved 11 March 2018
Attard, C., Ingram, N., Forgasz, H., Leder, G. & Grootenboer, P., 2016, 'Mathematics education and the affective domain', in K. Makar, S. Dole, J. Visnovska, M. Goos, A. Bennison & K. Fry (eds), *Research in Mathematics Education in Australasia, 2012–2015*, pp. 73–96, Singapore: Springer
Australian Association of Mathematics Teachers *see* AAMT
Australian Bureau of Statistics, 2012, *1301.0 Year Book Australia, Labour, Persons not in the labour force*, <www.abs.gov.au/ausstats/abs@.nsf/Lookup/by%20Subject/1301.0~2012~Main%20Features~Persons%20not%20in%20the%20labour%20force~298>, retrieved 10 August 2018
Australian Bureau of Statistics, 2014, *4228.0—Programme for the International Assessment of Adult Competencies, Australia, 2011–2012: Preliminary findings*, <www.abs.gov.au/ausstats/abs@.nsf/Lookup/4228.0main+features992011-2012>, retrieved 1 September 2017
Australian Bureau of Statistics, 2018a, *6102.0.55.001 Labour Statistics: Concepts, Sources and Methods, Feb 2018. The Labour Force Framework*, <www.abs.gov.au/ausstats/abs@.nsf/Lookup/by%20Subject/6102.0.55.001~Feb%202018~Main%20Features~The%20Labour%20Force%20Framework~3>, retrieved 10 August 2018
Australian Bureau of Statistics, 2018b, *6102.0.55.001 Labour Statistics: Concepts, Sources and Methods, Feb 2018. Not in the Labour Force*, <www.abs.gov.au/ausstats/abs@.nsf/Lookup/by%20Subject/6102.0.55.001~Feb%202018~Main%20Features~Not%20in%20the%20Labour%20Force~8>, retrieved 10 August 2018

Australian Bureau of Statistics, 2018c, *6226.0 Participation, Job Search and Mobility, Australia, February 2018. Key Findings*, <www.abs.gov.au/ausstats/abs@.nsf/mf/6226.0>, retrieved 10 August 2018

Australian Council for Educational Research *see* ACER

Australian Curriculum, Assessment and Reporting Authority *see* ACARA

Australian Education Council, 1989, *The Hobart Declaration on Schooling (1989)*, <www.educationcouncil.edu.au/EC-Publications/EC-Publications-archive/EC-The-Hobart-Declaration-on-Schooling-1989.aspx>, retrieved 9 July 2018

Australian Education Council, 1997, *National Report on Schooling in Australia*, <http://scseec.edu.au/archive/Publications/Publications-archive/National-Report-on-Schooling-in-Australia/ANR-1997.aspx>, retrieved 12 November 2017

Australian Institute for Teaching and School Leadership *see* AITSL

Barrington, F. & Evans, M., 2016, *Year 12 Mathematics Participation in Australia: The last ten years*, <http://amsi.org.au/wp-content/uploads/2016/09/barrington-2016.pdf>, retrieved 11 March 2018

Bennison, A., 2015, 'Developing an analytic lens for investigating identity as an embedder-of-numeracy', *Mathematics Education Research Journal*, vol. 27, pp. 1–19

Bennison, A., 2017, 'Re-examining a framework for teacher identity as an embedder-of-numeracy', in A. Downton, S. Livy & J. Hall (eds), *40 Years On: We are still learning*, Proceedings of the 40th annual conference of the Mathematics Education Research Group of Australasia, pp. 101–8, Melbourne: MERGA

Bessot, A., 1996, 'Geometry and work: Examples from the building industry', in A. Bessott & J. Ridgeway (eds), *Education for Mathematics in the Workplace*, pp. 143–58, Dordrecht: Kluwer

Blow, F., Lee, P. & Shemilt, D., 2012, 'Time and chronology: Conjoined twins or distant cousins?', *Teaching History*, vol. 147, pp. 26–34

Board of Teacher Registration, Queensland, 2005, *Numeracy in Teacher Education: The way forward in the 21st century*, <http://qct.edu.au/pdf/Archive/BTR_NumeracyReport2005.pdf>, retrieved 17 August 2017

Burkhart, H. & Swan, M., 2013, 'Task design for systemic improvement: Principles and frameworks', in C. Margolinas (ed.), *Task Design in Mathematics Education*, Proceedings of ICMI Study 22, July 2014, Oxford, pp. 431–9, Oxford: ICMI, <https://hal.archives-ouvertes.fr/hal-00834054/file/ICMI_STudy_22_proceedings_2013-10.pdf>, retrieved 4 March 2018

Carter, M., 2015, 'A multiple case study of NAPLAN numeracy testing of Year 9 students in three Queensland secondary schools', unpublished PhD thesis, <https://eprints.qut.edu.au/79906/>, retrieved 8 August 2018

Carter, M., Klenowski, V. & Chalmers, C., 2015, 'Challenges in embedding numeracy throughout the curriculum in three Queensland secondary schools', *Australian Educational Researcher*, vol. 42, pp. 595–611

Chen, L., Bae, S.R., Battista, C., Qin, S., Chen, T., Evans, T.M. & Menon, V., 2018, 'Positive attitude toward math supports early academic success: Behavioral evidence and neurocognitive mechanisms', *Psychological Science*, vol. 29(3), pp. 390–402

City, E.A., Elmore, R.F., Fiarman, S.E. & Teitel, L., 2009, *Instructional Rounds in Education*, Cambridge, MA.: Harvard Educational Press

Coben, D. & Weeks, K., 2014, 'Meeting the mathematical demands of the safety-critical workplace: Medication dosage calculation problem-solving for nursing', *Educational Studies in Mathematics*, vol. 86, pp. 253–70

Cockcroft, W., 1982, *Mathematics Counts*, London: HMSO

Cooper, C., Dole, S., Geiger, V. & Goos, M., 2012, 'Numeracy in society and environment', *Australian Mathematics Teacher*, vol. 68, no. 1, pp. 16–20

Council of Australian Governments, 2008, *National Numeracy Review Report*, <http://webarchive.nla.gov.au/gov/20080718164654/http://www.coag.gov.au/reports/index.htm#numeracy>, retrieved 11 March 2018

Crowe, A., 2010, '"What's math got to do with it?" Numeracy and social studies education', *The Social Studies*, vol. 101, no. 3, pp. 105–10, doi: 10.1080/00377990903493846

Cumming, J., 1996, *Adult Numeracy Policy and Research in Australia: The present context and future directions*, <http://files.eric.ed.gov/fulltext/ED405485.pdf>, retrieved 20 October 2017

DEETYA (Department of Employment, Education, Training and Youth Affairs), 1997, *Numeracy = Everyone's Business: The report of the Numeracy Education Strategy Development Conference, May 1997*, Adelaide: Australian Association of Mathematics Teachers

Department of Education and Children's Services, 2005, *South Australian Curriculum, Standards and Accountability Framework: Middle years band*, <www.sacsa.sa.edu.au/ATT/%7B85CFF734-68DE-4F6D-A626-4EA1EDEC69C2%7D/SACSA_6_MYB.pdf>, retrieved 5 May 2017

Department of Education and Children's Services, 2009, *Numeracy in the Middle Years Curriculum: A resource paper; An audit of numeracy in the SACSA Framework*, <https://numeracy4schools.files.wordpress.com/2015/03/numeracy-audit-book_v6.pdf>, retrieved 12 March 2018

Department of Education, Science and Training *see* DEST

Department of Education, Training and Youth Affairs, 2000, *Numeracy, a Priority for All: Challenges for Australian schools*, Canberra: DETYA

Department of Employment, Education, Training and Youth Affairs *see* DEETYA

DEST (Department of Education, Science and Training), 2004a, *Numeracy Across the Curriculum*, Canberra: Commonwealth of Australia

DEST (Department of Education, Science and Training), 2004b, *Numeracy Across the Curriculum: Appendices*, Canberra: Commonwealth of Australia

DEST (Department of Education, Science and Training), 2005, *Numeracy Research and Development Initiative, 2001–2004: An overview of the numeracy projects*, Canberra: Commonwealth of Australia

Diezmann, C., Watters, J. & English, L., 2001, 'Implementing mathematical investigations with young children', in J. Bobis, B. Perry & M. Mitchelmore (eds), *Numeracy and Beyond*, Proceedings of the 24th annual conference of the Mathematics Education Research Group of Australasia, pp. 170–7, Sydney: MERGA

Doig, B., 2001, *Summing Up: Australian numeracy performances, practices, programs and possibilities*, Camberwell, Vic.: Australian Council for Educational Research

Drijvers, P. & Weigand, H., 2010, 'The role of handheld technology in the mathematics classroom', *ZDM: The International Journal on Mathematics Education*, vol. 42, no. 7, pp. 665–6

Education Council, 2015, *National STEM School Education Strategy, 2016–2026*, <www.educationcouncil.edu.au/site/DefaultSite/filesystem/documents/National%20STEM%20School%20Education%20Strategy.pdf>, retrieved 11 March 2018

Education Queensland, 2005, *Professional Standards for Teachers: Guidelines for professional practice*, <http://education.qld.gov.au/staff/development/pdfs/profstandards.pdf>, retrieved 26 March 2018

Enyedy, N., Goldberg, J. & Welsh, K.M., 2005, 'Complex dilemmas of identity and practice', *Science Education*, vol. 90, pp. 68–93, doi: 10.1002/sce.20096

Ernest, P., 2002, 'Empowerment in mathematics education', *Philosophy of Mathematics Journal*, vol. 15 <www.ex.ac.uk/~PErnest/pome15/contents.htm>, retrieved 9 August 2018

Forgasz, H. & Hall, J., 2016, 'Numeracy for learners and teachers: Evaluation of an MTeach coursework unit at Monash University', in B. White, M. Chinnappan & S. Trenholm, S. (eds), *Opening Up Mathematics Education Research*, Proceedings of the 39th annual conference of the Mathematics Education Research Group of Australasia, pp. 228–35, Adelaide: MERGA

Gapminder, 2018a, bubbles graph showing CO_2 emissions (per capita) over time, <http://bit.ly/2mFzmhN>, retrieved 25 July 2018

Gapminder, 2018b, bubbles graph showing cumulative CO_2 emissions over time, <http://bit.ly/2OfqN9W>, retrieved 25 July 2018

Gee, J.P., 2001, 'Identity as an analytic lens for research in education', *Review of Research in Education*, vol. 25, pp. 99–125

Geiger, V., 2016, 'Teachers as designers of effective numeracy tasks', in B. White, M. Chinnappan & S. Trenholm (eds), *Opening Up Mathematics Education Research*, Proceedings of the 39th annual conference of the Mathematics Education Research Group of Australasia, pp. 252–9, Adelaide: MERGA

Geiger, V., Forgasz, H. & Goos, M., 2015, 'A critical orientation to numeracy across the curriculum', *ZDM Mathematics Education*, vol. 47, no. 4, pp. 611–24, doi: 10.1007/s11858/014-0648-1

Geiger, V., Goos, M. & Dole, S., 2013, 'Taking advantage of incidental school events to engage with the applications of mathematics: The case of surviving the reconstruction', in G. Stillman, G. Kaiser, W. Blum & J. Brown (eds), *Teaching Mathematical Modelling: Connecting to research and practice*, pp. 175–84, Dordrecht: Springer

Geiger, V., Goos, M. & Dole, S., 2014, 'Curriculum intent, teacher professional development and student learning in numeracy', in Y. Li & G. Lappan (eds), *Mathematics Curriculum in School Education*, pp. 473–92, New York: Springer

Geiger, V., Goos, M. & Dole, S., 2015, 'The role of digital technologies in numeracy teaching and learning', *International Journal of Science and Mathematics Education*, vol. 13, no. 5, pp. 1115–37, doi: 10.1007/s10763-014-9530-4

Geiger, V., Goos, M., Dole, S., Forgasz, H. & Bennison, A., 2013, 'Exploring the demands and opportunities for numeracy in the Australian Curriculum: English', in V. Steinle, L. Ball & C. Bardini (eds), *Mathematics Education: Yesterday, today and tomorrow*, Proceedings of the 36th annual conference of the Mathematics Education Research Group of Australasia, vol. 1, pp. 330–7, Melbourne: MERGA

Geiger, V., Goos, M., Dole, S., Forgasz, H. & Bennison, A., 2014, 'Devising principles of design for numeracy tasks', in J. Anderson, M. Cavanagh & A. Prescott (eds), *Curriculum in Focus: Research guided practice*, Proceedings of the 37th annual conference of the Mathematics

Education Research Group of Australasia, pp. 239–46, Sydney: MERGA

Geiger, V., Goos, M. & Forgasz, H., 2015, 'A rich interpretation of numeracy for the 21st century: A survey of the state of the field', *ZDM Mathematics Education*, vol. 47, no. 4, pp. 531–48

Gibbs, M., Goos, M., Geiger, V. & Dole, S., 2012, 'Numeracy in secondary school mathematics', *Australian Mathematics Teacher*, vol. 68, no. 1, pp. 29–35

Goodnough, K., 2011, 'Examining the long-term impact of collaborative action research on teacher identity and practice: The perceptions of K-12 teachers', *Educational Action Research*, vol. 19, no. 1, pp. 73–86, doi: 10.1080/09650792.2011.547694

Goos, M., Dole, S. & Geiger, V., 2012, 'Auditing the numeracy demands of the Australian Curriculum', in J. Dindyal, L. Chen & S.F. Ng (eds), *Mathematics Education: Expanding horizons*, Proceedings of the 35th annual conference of the Mathematics Education Research Group of Australasia, pp. 314–21, Singapore: MERGA

Goos, M., Geiger, V., Bennison, A. & Roberts, J., 2015, *Numeracy Teaching Across the Curriculum in Queensland: Resources for teachers; Final report*, <http://qct.edu.au/pdf/Numeracy_Teaching_Across_Curriculum_QLD.pdf>, retrieved 11 March 2018

Goos, M., Geiger, V. & Dole, S., 2010, 'Auditing the numeracy demands of the middle years curriculum', in L. Sparrow, B. Kissane & C. Hurst (eds), *Shaping the Future of Mathematics Education*, Proceedings of the 33rd annual conference of the Mathematics Education Research Group of Australasia, pp. 210–17, Fremantle: MERGA

Goos, M., Geiger, V. & Dole, S., 2011, 'Teachers' personal conceptions of numeracy', in B. Ubuz (ed.), *Proceedings of the 35th Conference of the International Group for the Psychology of Mathematics Education*, vol. 2, pp. 457–64, Ankara: PME

Goos, M., Geiger, V. & Dole, S., 2013, 'Designing rich numeracy tasks', in C. Margolinas (ed.), *Task Design in Mathematics Education*, Proceedings of ICMI Study 22, July 2014, Oxford, pp. 589–98, Oxford: ICMI, <https://hal.archives-ouvertes.fr/hal-00834054/file/ICMI_STudy_22_proceedings_2013-10.pdf>, retrieved 4 March 2018

Goos, M., Geiger, V. & Dole, S., 2014, 'Transforming professional practice in numeracy teaching', in Y. Li, E. Silver & S. Li (eds), *Transforming Mathematics Instruction: Multiple approaches and practices*, pp. 81–102, New York: Springer

Graven, M., 2004, 'Investigating mathematics teachers' learning within an in-service community of practice: The centrality of confidence', *Educational Studies in Mathematics*, vol. 57, pp. 177–211

Hart, A., 2016, 'Introducing the new, realistic Barbie: "The thigh gap has officially gone"', *Telegraph*, 28 January, <www.telegraph.co.uk/news/shopping-and-consumer-news/12122027/Introducing-the-new-realistic-Barbie-The-thigh-gap-has-officially-gone.html>, retrieved 9 July 2018

Hiebert, J. & Grouws, D.A., 2007, 'The effects of classroom mathematics teaching on students' learning', in F.K. Lester (ed.), *Second Handbook of Research on Mathematics Teaching and Learning*, pp. 371–404, Charlotte, NC: National Council of Teachers of Mathematics

Hogan, J., 2000a, 'Numeracy: Across the curriculum?', *Australian Mathematics Teacher*, vol. 56, no. 3, pp. 17–20

Hogan, J., 2000b, *The Numeracy Audit: An overview*, <http://redgumconsulting.com.au/Numeracy%20Audit.ps1.pdf>, retrieved 9 July 2018

Hogan, J., 2002, *The Numeracy Research Circle*, <www.redgumconsulting.com.au/num_research.html>, retrieved 9 July 2018

Hoyles, C., Noss, R., Kent, P. & Bakker, A., 2010, *Improving Mathematics at Work: The need for techno-mathematical literacies*, London: Routledge

Hoyles, C., Wolf, A., Molyneux-Hodgson, S. & Kent, P., 2002, *Mathematical skills in the workplace: Final report to the Science, Technology and Mathematics Council*, London: Institute of Education, University of London; Science, Technology and Mathematics Council, <http://discovery.ucl.ac.uk/10001565/1/Hoyles2002MathematicalSkills.pdf>, retrieved 9 August 2018

Hughes-Hallett, D., 2001, 'Achieving numeracy: The challenge of implementation', in L. Steen (ed.), *Mathematics and Democracy: The case for quantitative literacy*, pp. 93–8, Princeton, NJ: National Council on Education and the Disciplines

International Association for the Evaluation of Educational Achievement, 2018, *TIMSS: Trends in International Mathematics and Science Study*, <www.iea.nl/timss>, retrieved 9 July 2018

Jablonka, E., 2015, 'The evolvement of numeracy and mathematical literacy curricula and the construction of hierarchies of numerate or mathematically literate subjects', *ZDM Mathematics Education*, vol. 47, no. 4, pp. 599–609

Jorgensen Zevenbergen, R., 2011, 'Young workers and their dispositions towards mathematics: Tensions of a mathematical habitus in the retail industry', *Educational Studies in Mathematics*, vol. 76, pp. 87–100

Kanes, C., 1996, 'Investigating the use of language and mathematics in the workplace setting', in P. Clarkson (ed.), *Technology in Mathematics Education*, Proceedings of the 19th annual conference of the Mathematics Education Research Group of Australasia, pp. 314–21, Melbourne: MERGA

Kemp, M. & Hogan, J., 2000, *Planning for an Emphasis on Numeracy in the Curriculum*, <www.aamt.edu.au/content/download/1251/25266/file/kemp-hog.pdf>, retrieved 27 September 2016

Kieran, C., Doorman, M. & Ohtani, M., 2013, 'Principles and frameworks for task design within and across communities', in C. Margolinas (ed.), *Task Design in Mathematics Education*, Proceedings of ICMI Study 22, July 2014, Oxford, pp. 419–20, Oxford: ICMI, <https://hal.archives-ouvertes.fr/hal-00834054v3/document>, retrieved 17 November 2017

Lamb, S., 1997, *School Achievement and Initial Education and Labour Market Outcomes*, LSAY Research Report No. 4, <www.lsay.edu.au>, retrieved 12 March 2018

Lappan, G. & Phillips, E., 2009, 'A designer speaks', *Educational Designer*, vol. 1, no. 3, <www.educationaldesigner.org/ed/volume1/issue3/>, retrieved 9 August 2018

Leder, G.C. & Forgasz, H.J., 2002, 'Measuring mathematical beliefs and their impact on the learning of mathematics: A new approach', in G. Leder, E. Pehkonen & G. Toerner (eds), *Beliefs: A hidden variable in mathematics education?* (pp. 95–113). Dordrecht: Kluwer

Maass, K., Garcia, J., Mousoulides, N. & Wake, G., 2013, 'Designing interdisciplinary tasks in an international design community', in C. Margolinas (ed.), *Task Design in Mathematics Education*, Proceedings of ICMI Study 22, July 2014, Oxford, pp. 367–76, Oxford: ICMI, <https://hal.archives-ouvertes.fr/hal-00834054/file/ICMI_STudy_22_proceedings_2013-10.pdf>, retrieved 4 March 2018

MCEETYA (Ministerial Council on Education, Employment, Training and Youth Affairs), 1999, *The Adelaide Declaration of National Goals for Schooling in the Twenty-First Century*, <www.scseec.edu.au/archive/Publications/Publications-archive/The-Adelaide-Declaration.aspx>, retrieved 9 August 2018

MCEETYA (Ministerial Council on Education, Employment, Training and Youth Affairs), 2008, *Melbourne Declaration on Educational Goals for Young Australians*, <www.curriculum.edu.au/verve/_resources/National_Declaration_on_the_Educational_Goals_for_Young_Australians.pdf>, retrieved 17 November 2017

McLeod, D.B., 1992, 'Research on affect in mathematics education: A reconceptualization', in D.A. Grouws (ed.), *Handbook of Research on Mathematics Teaching and Learning*, pp. 575–96, New York: Macmillan

Messerli, F., 2012, 'Chocolate consumption, cognitive function, and Nobel Laureates', *New England Journal of Medicine*, vol. 367, no. 16, pp. 1562–4

Milton, M., Rohl, M. & House, H., 2007, 'Secondary beginning teachers' preparedness to teach literacy and numeracy: A survey', *Australian Journal of Teacher Education*, vol. 32, no. 2, <http://ro.ecu.edu.au/ajte/vol32/iss2/4>, retrieved 9 August 2018

Ministerial Council on Education, Employment, Training and Youth Affairs *see* MCEETYA

Ministry of Education, 1959, *15 to 18: A report of the Central Advisory Council for Education*, London: HMSO

Morony, W., Hogan, J. & Thornton, S., 2004, 'Numeracy across the curriculum', *Australian National Schools Network Snapshot*, vol. 1, pp. 1–12

NAP (National Assessment Program), 2016a, *How to Interpret*, <www.nap.edu.au/results-and-reports/how-to-interpret>, retrieved 6 July 2018

NAP (National Assessment Program), 2016b, *National Reports*, <www.nap.edu.au/results-and-reports/national-reports>, retrieved 6 July 2018

NAP (National Assessment Program), 2016c, *Numeracy*, <www.nap.edu.au/naplan/numeracy>, retrieved 11 May 2018

NAP (National Assessment Program), 2016d, *Scales*, <www.nap.edu.au/results-and-reports/how-to-interpret/scales>, retrieved 11 May 2018

NAP (National Assessment Program), 2016e, *Score Equivalence Tables*, <www.nap.edu.au/results-and-reports/how-to-interpret/score-equivalence-tables>, retrieved 6 July 2018

NAP (National Assessment Program), 2016f, *Standards*, <www.nap.edu.au/results-and-reports/how-to-interpret/standards>, retrieved 6 July 2018

NAP (National Assessment Program), 2016g, *Student Reports*, <www.nap.edu.au/results-and-reports/student-reports>, retrieved 6 July 2018

NAP (National Assessment Program) & ACARA (Australian Curriculum, Assessment and Reporting Authority), 2017, *Numeracy: Calculator allowed; Year 7 example test*, <www.nap.edu.au/docs/default-source/default-document-library/nap17_y7_numeracy_ca_exampletest.pdf?sfvrsn=2>, retrieved 6 July 2018

National Assessment Program *see* NAP

National Curriculum Board, 2008, *National Mathematics Curriculum: Framing paper*, November, <https://acaraweb.blob.core.windows.net/

resources/National_Mathematics_Curriculum_-_Framing_Paper.pdf>, retrieved 11 March 2018

National Curriculum Board, 2009, *Shape of the Australian Curriculum: Mathematics*, May, <https://acaraweb.blob.core.windows.net/resources/The_Shape_of_the_Australian_Curriculum_May_2009_file.pdf>, retrieved 11 March 2018

New London Group, 1996, 'A pedagogy of multiliteracies: Designing social futures', *Harvard Educational Review*, vol. 66, no. 1, pp. 60–93

Noss, R., 1998, 'New numeracies of a technological culture', *For the Learning of Mathematics*, vol. 18, no. 2, pp. 2–12

Noss, R., Hoyles, C. & Pozzi, S., 2000, 'Working knowledge: Mathematics in use', in A. Bessot & J. Ridgeway (eds), *Education for Mathematics in the Workplace*, pp. 17–35, Dordrecht: Kluwer

OECD (Organisation for Economic Co-operation and Development), 2012a, *Literacy, Numeracy and Problem Solving in Technology-Rich Environments: Framework for the OECD Survey of Adult Skills*, <http://dx.doi.org/10.1787/9789264128859-en>, retrieved 11 March 2018

OECD (Organisation for Economic Co-operation and Development), 2012b, *Released Mathematics Questions*, <www.oecd.org/pisa/test/PISA%202012%20items%20for%20release_ENGLISH.pdf>, retrieved 6 July 2018

OECD (Organisation for Economic Co-operation and Development), 2016, *PISA 2015 Assessment and Analytical Framework: Science, reading, mathematics and financial literacy*, Paris: OECD

OECD (Organisation for Economic Co-operation and Development), 2018a, *Skills Matter*, <www.oecd.org/skills/piaac/skills-matter-9789264258051-en.htm>, retrieved 6 July 2018

OECD (Organisation for Economic Co-operation and Development), 2018b, *Survey of Adult Skills*, <www.oecd.org/skills/piaac/>, retrieved 6 July 2018

OECD (Organisation for Economic Co-operation and Development), 2018c, *Survey of Adult Skills: First results; Australia, Country note*, <www.oecd.org/skills/piaac/Country%20note%20-%20Australia_final.pdf>, retrieved 6 July 2018

Office of the Chief Scientist, 2014, *Science, Technology, Engineering and Mathematics: Australia's future*, <www.chiefscientist.gov.au/wp-content/uploads/STEM_AustraliasFuture_Sept2014_Web.pdf>, retrieved 11 March 2018

Organisation for Economic Co-operation and Development *see* OECD

Parsons, S. & Bynner, J., 2005, *Does Numeracy Matter More?*, London: National Research and Development Centre for Adult Literacy and Numeracy

Peters, C., Geiger, V., Goos, M. & Dole, S., 2012, 'Numeracy in health and physical education', *Australian Mathematics Teacher*, vol. 68, no. 1, pp. 21–7

Philipp, R., 2007, 'Mathematics teachers' beliefs and affect', in F.K. Lester, Jnr (ed.), *Second Handbook on Mathematics Teaching and Learning*, pp. 257–315, Charlotte, NC: National Council of Teachers of Mathematics

Phillips, I., 2002, 'History and mathematics or history with mathematics: Does it add up?', *Teaching History*, vol. 107, pp. 35–40

Quantitative Literacy Design Team, 2001, 'The case for quantitative literacy', in L. Steen (ed.), *Mathematics and Democracy: The case for quantitative literacy*, pp. 1–22, Princeton, NJ: National Council on Education and the Disciplines

Rosa, M. & Clark Orey, D., 2015, 'A trivium curriculum for mathematics based on literacy, matheracy, and technoracy: An ethnomathematical perspective', *ZDM Mathematics Education*, vol. 47, no. 4, pp. 587–98

Rosling, H., 2011, *Florence Nightingale: Joy of stats (3/6)*, <www.youtube.com/watch?v=yhX0OR1_Vfc>, retrieved 9 July 2018

Roth, R., 2011, 'So fast', <https://goodmenproject.com/dadsgood-2/so-fast/>, retrieved 7 August 2018

Sachs, J., 2005, 'Teacher education and the development of professional identity', in P. Denicolo & M. Kompf (eds), *Connecting Policy and Practice: Challenges for teaching and learning in schools and universities*, pp. 5–21, New York: Routledge Farmer

Sfard, A. & Prusak, A., 2005, 'Telling identities: In search of an analytic tool for investigating learning as a culturally shaped activity', *Educational Researcher*, vol. 34, no. 4, pp. 14–22

Shulman, L.S., 1987, 'Knowledge and teaching: Foundations of the new reform', *Harvard Educational Review*, vol. 57, no. 1, pp. 1–21

Straesser, R., 2007, 'Didactics of mathematics: More than mathematics and school!' *ZDM: The International Journal on Mathematics Education*, vol. 39, no. 1, pp. 165–71

Sullivan, P., 2011, *Teaching Mathematics: Using research-informed strategies*, Australian Education Review no. 59, Camberwell, Vic.: Australian Council for Educational Research

Sullivan, P., Clarke, D. & Clarke, B., 2013, *Teaching with Tasks for Effective Mathematics Learning*, New York: Springer

Sullivan, P. & Yang, Y., 2013, 'Features of task design informing teachers' decisions about goals and pedagogies', in C. Margolinas (ed.), *Task Design in Mathematics Education*, Proceedings of ICMI Study 22, July 2014, Oxford, pp. 529–30, Oxford: ICMI, <https://hal.archives-ouvertes.fr/hal-00834054/file/ICMI_STudy_22_proceedings_2013-10.pdf>, retrieved 4 March 2018

Teacher Education Ministerial Advisory Group, 2014, *Action Now: Classroom ready teachers*, <https://docs.education.gov.au/node/36783>, retrieved 23 April 2018

Thomson, S., De Bortoli, L. & Underwood, C., 2017, *PISA 2015: Reporting Australia's results*, <http://research.acer.edu.au/cgi/viewcontent.cgi?article=1023&context=ozpisa>, retrieved 11 March 2018

Thomson, S., Lokan, J., Lamb, S. & Ainley, J., 2003, *Lessons from the Third International Mathematics and Science Study*, <https://research.acer.edu.au/timss_monographs/9/>, retrieved 9 August 2018

Thornton, S. & Hogan, J., 2003, 'Numeracy across the curriculum: Demands and opportunities', paper presented at the annual conference of the Australian Curriculum Studies Association, Adelaide, 28–30 September, <www.acsa.edu.au/pages/images/thornton_-_numeracy_across_the_curriculum.pdf>, retrieved 18 August 2017

Thornton, S. & Hogan, J., 2004, 'Orientations to numeracy: Teachers' confidence and disposition to use mathematics across the curriculum', in M. Johnson Hoines & A. Berit Fugelstad (eds), *Proceedings of the 28th Conference of the International Group for the Psychology of Mathematics Education*, vol. 4, pp. 313–20, Bergen: PME

University of Queensland, 2018, *The University of Queensland*, <www.filmpond.com/ponds/qct-the-university-of-queensland>, retrieved 9 July 2018

Van Zoest, L. & Bohl, J., 2005, 'Mathematics teacher identity: A framework for understanding secondary school mathematics teachers' learning through practice', *Teacher Development*, vol. 9, pp. 315–45, doi: 10.1080/13664530500200258

Wake, G., 2014, 'Making sense of and with mathematics: The interface between academic mathematic and mathematics in practice', *Educational Studies in Mathematics*, vol. 86, pp. 271–90

Wenger, E., 1998, *Communities of Practice: Learning, meaning and identity*, Cambridge, UK: Cambridge University Press

Wilkins, J. & Hicks, D., 2001, 'A s(t)imulating study of map projections: An exploration integrating mathematics with social studies', *Mathematics Teacher*, vol. 94, no. 8, pp. 660–5

Willis, K., Geiger, V., Goos, M. & Dole, S., 2012, 'Numeracy for what's in the news and building an expressway', *Australian Mathematics Teacher*, vol. 68, no. 1, pp. 9–15

Willis, S., 1990a, *Being Numerate: Whose right, who's left?*, Proceedings of the Australian Council for Adult Literacy conference The Right to Literacy: The rhetoric, the romance, the reality, vol. 1, pp. 77–94, Sydney: NSW Adult Literacy and Numeracy Council, <http://files.eric.ed.gov/fulltext/ED367779.pdf#page=80>, retrieved 18 October 2017

Willis, S., 1990b, *Being Numerate: What counts?*, Hawthorn, Vic: Australian Council for Education Research

Willis, S., 1998, 'Which numeracy?', *Unicorn*, vol. 24, no. 2, pp. 32–42

Willis, S., 2001, 'Developing a mathematics curriculum for the compulsory years of schooling: Some issues', *Mathematics K-10 Symposium Proceedings*, pp. 8–18, Sydney: Board of Studies NSW, <http://citeseerx.ist.psu.edu/viewdoc/download?doi=10.1.1.203.3173&rep=rep1&type=pdf>, retrieved 12 March 2018

Wilson, S. & Thornton, S., 2006, 'To heal and enthuse: Developmental bibliotherapy and pre-service primary teachers' reflections on learning and teaching mathematics', in P. Grootenboer, R. Zevenbergen & M. Chinnappan (eds), *Identities, Cultures and Learning Spaces*, Proceedings of the 29th annual conference of the Mathematics Education Research Group of Australasia, pp. 35–44, Canberra: MERGA

Yes (Prime) Minister Files, 2006, *YPM 1.1: The Grand Design*, <www.yes-minister.com/ypmseas1a.htm>, retrieved 6 July 2018

Zevenbergen, R., 1995, 'Towards a socially critical numeracy', *Critical Forum*, vol. 3, nos 2–3, pp. 82–102

INDEX

21st Century Numeracy
    Model 58–77, 108, 126,
    136, 178, 191, 209, 216,
    227
  assessing numeracy 197–8
  case studies 70
  contexts 58
  critical orientation 58, 66–9
  definition of numeracy 179
  dimensions 58, 59
  mathematical knowledge 58,
    61–2
  positive dispositions 58, 63
  strengths and limitations
    76–7
  tools 58, 64–5
  unemployment case study 70,
    71–3

Adelaide Declaration of
    National Goals for
    Schooling in the
    Twenty-First Century 34
affective domain 221
arts 213
  analysis and response 88
  auditing in 86–8
  contexts, in 88
  practice 87
assessment *see* numeracy
    assessment
auditing
  numeracy audit *see* numeracy
    audit

numeracy demands *see*
    numeracy demands
Australia
  numeracy in 38–43
Australian Association of
    Mathematics Teachers 41
Australian Council for
    Educational Research 186
Australian Curriculum 50, 206,
    226
  developing 52
  disciplinary boundaries
    212–13
  Foundation to Year 10 52
  learning areas in 212
  numeracy in 52, 80–2, 177–8,
    179, 193–4
    audit process 85
    fingerprints 92
    icons and filters 81–2
    six key ideas 178
    STEM education 214
Australian Curriculum:
    Mathematics 51, 179, 180,
    206
Australian Institute for Teaching
    and School Leadership
    53–4, 192, 193, 226, 227
Australian Professional
    Standards for Teachers 22,
    50, 53–5, 76, 157–8, 192–3,
    226
  focus areas 54
  Standard 2 77

calculators 10
  reliance on 1
cartograms 15
  comparisons using 16
challenge, importance of 130–1
chocolate bar case study 73–5
citizenship, critical 13
civic life, numeracy in 13
civics education 77
classroom
  numeracy development 191
  observations 210
Cockcroft Report 2, 37
community, numeracy in 13
Connected Mathematics 129
content knowledge 104
context(s)
  21st Century Numeracy
    Model 58
  arts in 88
  mathematics, auditing in 99
  SACSA Framework 89–91,
    94
context domain 222–3
Council of Australian
    Governments 51
'critic' role 45
critical and creative
    thinking 213
critical orientation 198, 210,
    211
  21st Century Numeracy
    Model 58, 66–9
  SACSA Framework 92, 96
cross-curricular numeracy 51,
    106–8, 152–3
  Australian Curriculum 153
  challenges to 212

initial teacher education,
    in 226
  mathematics and 96
  planning for 128
  students' perspectives on
    223–6
Crowther Report 2, 37
'cultural numeracy' 40
curriculum
  Australian see Australian
    Curriculum
  changes in 10
  cross-curricular numeracy
    see cross-curricular
    numeracy
  numeracy in 21–5

Department of Education
    Queensland
  Literacy and Numeracy
    Strategy 1994–98 42
Department of Education,
    Training and Youth
    Affairs 42, 152
Department of Employment,
    Education, Training and
    Youth Affairs 41
diagrams 15
digital technologies 6, 40, 210
digital tools 64, 91, 100, 222
dispositions 198, 199
  21st Century Numeracy
    Model 63
  assessing 199
  mathematics, auditing in 99
  positive 4, 63
  SACSA Framework 91, 95
domains of influence 218

educational policy, numeracy in 46–50
employment
  numeracy and 20
  unemployment case study 70, 71–3
English 77, 212
  Australian Curriculum 110
  numeracy in 110
  numeracy task to support learning in 144
environmental education 77

flip 140
'fluent operator' role 45
'functional numeracy' 39

geography 79
globalisation 40
'graduate teacher' 53
graphs 15

health and physical education (HPE) 77, 213
  Australian Curriculum 104
  numeracy 108
  pedometer case study 109
  Years 7 and 8 Curriculum 82, 108
'highly accomplished teacher' 53
history 79
  Australian Curriculum 112
  historical concepts 113
  historical skills 119–20
  numeracy in 112–14
Hobart Declaration on Schooling 34

homes, numeracy in 6
HPE *see* health and physical education (HPE)
humanities 212

information and communication technology 35, 213
internet 10

know-how 61
  numerate, for being 45
  strategic 45
knowledge domain 220

languages 213
'lead teacher' 53
'learner' role 45
learning *see also* numeracy learning
  areas 213
  improving 130
life history domain 218–20
literacy 20, 213
  definitions 40
  goals for 42
  life outcomes and 21
  teachers incorporating 23–4
Literacy and Numeracy Test for Initial Teacher Education Students (LANTITE) 179, 188
literate
  definitions 40
longitudinal studies
  Australia, in 20
  United Kingdom, in 20
Longitudinal Surveys of Australian Youth 20

maps 15
mathematical empowerment 67
mathematical investigations 130
mathematical knowledge 4,
    61–2, 88, 97, 198, 211, 214
  arts, auditing in 92–4
  demands 85
  mathematics, auditing in
    98–9
  SACSA Framework 92–4
  workplace, in 60
mathematical literacy 4
  decline in 185, 214
  definition 5, 185, 189
mathematical skills 3, 4
  assessing 198
mathematicians
  professional, work of 18
mathematics 21, 51, 52, 212
  auditing numeracy demands
    in 98
  contexts 99
  critical orientation 100
  critique, tool for 40
  cross-curricular numeracy
    and 96–8
  definition 18
  dispositions of students
    99–100
  everyday life, in 4
  numeracy, distinguished
    18–19, 44, 51
  real world, in 59–63
  social empowerment, role
    in 66
  specialist teachers, role of 98
  tools 100
  utilitarian purpose 66

Melbourne Declaration on
    Educational Goals for Young
    Australians 35
Mercator projection 15
Ministerial Council for
    Education, Early Childhood
    Development and Youth
    Affairs (MCEEDYA) 53
Ministerial Council on
    Education, Employment,
    Training and Youth Affairs
    (MCEETYA) 41
multiliteracies 40
My School website 182

National Assessment Program—
    Literacy and Numeracy
    (NAPLAN) 35, 51, 153,
    179–80
  Catholic school 154
  government high school
    results 154
  implied definition of
    numeracy 179, 180
  independent school 154
  numeracy test in 180
  results 182
National Numeracy
    Review 50–2
Nightingale, Florence 194
numeracy
  21st century, in 57
  Australia, in 38–43
  communicative competence,
    as 44
  context, role of 3
  cross-curricular *see*
    cross-curricular numeracy

'cultural numeracy' 40
curriculum, in 21–25
definition 1–2, 6, 36–44, 51, 177, 179, 189–90
educational policy, in 45–50
'elastic term', as 40
embedders of 216
employment and 20
empowerment, as 40
enhancing 208
finding 6
'functional numeracy' 39
fundamental principles 50
goals for 42
historical development 33–56
home, in 6
importance of 19–21
introduction of term 2
learning area 213
life outcomes and 21
mathematics,
    distinguished 18–19, 44, 51
meaning 1–2
national goal of schooling 34
personal conception of 2
poor 20
problem-solving and 3
quantitative literacy 1, 4
reductionist view of 35
research and
    development 43–6
'socially critical numeracy' 40
strategic mathematics, as 44
teachers and *see* teacher(s)
United Kingdom, in 37–8
whole-school *see* whole-school numeracy

Numeracy Across the
    Curriculum Project 43–5
numeracy assessment 177–8, 206
    21st Century Numeracy Model 197
    classroom numeracy development 191
    formal 198
    informal 198
    national and international 179
numeracy audit 165–7
numeracy demands 79, 208
    arts, auditing in 86–8
    auditing 85–6
    mathematics, auditing in 98
numeracy education
    cross-curricular responsibility, as 21–5
    'embedder' 23
    'separatist' 22
    'theme-maker' 22
Numeracy Education
    Strategy Development
    Conference 41, 42, 50, 152
numeracy fingerprints 79, 85
    school curriculum, in 92
numeracy learning 180
    challenges in enhancing 208
    promoting 208, 218
    resources for 215
numeracy model 212
numeracy moment 117, 123
numeracy opportunities 79, 103, 124, 208
    curriculum goals 123
    promoting 123

Numeracy Research and
    Development Initiative 43
Numeracy Research Circle 163
numeracy skills 35
numeracy task design 131, 215
  actualising task 131, 140
  idea, identifying 131
  looking, noticing and
    seeing 132
  shaping idea into task 131,
    136
numerate 67
  attributes of being 38, 39, 42,
    43, 45, 213
  introduction of term 37
  know-how for 45

opinion piece, writing 111
Organisation for Economic
    Co-operation and
    Development (OECD)
    5, 20, 184

pedagogical architecture of
    lessons 140–1
pedagogy 104, 128
  choice of 130, 140
  investigative 130
pedometer case study 109
poverty, absolute 16, 17
problem-solving 3, 10, 19, 64
  strategies 3, 4, 61
professional attributes 24, 157
professional development
  model 160–2
  package 158
  projects 23
professional engagement 53

professional journals 158–9
professional knowledge 24, 53,
    156, 157
professional practice 25, 53, 157
'proficient teacher' 53
Programme for International
    Student Assessment
    (PISA) 5, 179, 184–6
Programme for the International
    Assessment of Adult
    Competencies (PIAAC) 20,
    179, 189
  numeracy results 191

quantitative literacy
  definition 4
  elements of 4–5
  numeracy as 1, 4
Quantitative Literacy Design
    Team 4, 19, 21, 51, 67
Queensland College of
    Teachers 156

research and development
    43–6

schooling
  debate and discussion
    about 41
  goals for 34–5, 41
  literacy as goal of 41
  numeracy as goal of 34–5, 41
  'socially just' 34
science 79, 212
Scootle 158
social domain 221–2
social empowerment 67
'socially critical numeracy' 40

South Australian Curriculum,
    Standards and
    Accountability (SACSA)
    Framework 85
  contexts 94
  critical orientation 96
  dispositions 95
  mathematical knowledge 92–4
  tools 95–6
Spanish conquest
  Americas, of 112–13
  case study 114–19
spreadsheets 10, 210
STEM disciplines 214–15
strategic know-how 45
students
  perspectives on
      cross-curriculum
      numeracy 223–6
subject knowledge 220
survey questions 13, 227

task design 129
  accessibility 131
  actualising task 140
  challenge 130–1
  effective 212
  fit to circumstance of
      tasks 130, 136–7
  numeracy *see* numeracy task
      design
  structuring task 136–7
teacher(s)
  beginning 23
  career stages 53
  content knowledge 104
  cross-curriculum
      numeracy 226

education programs 104
identity, developing 216–18
literacy, incorporating 23–4
numeracy
  conceptions of 3
  incorporating 23–4
  personal conceptions of 209
  responsibility for
      developing 80
  orientations to numeracy 22
  pre-service education 23
  role of 10
  self-assessment surveys 24,
      26–31
Teacher Education Ministerial
    Advisory Group 188
teaching, resources for 215
technologies 10, 213
terminology 4–6, 179
timelines
  case studies 120
  constructing 121
  features of 120
tools 198, 210
  21st Century Numeracy
      Model 64–5
  assessing use of 198
  mathematics, auditing in 100
  SACSA Framework 91, 95
Trends in International
    Mathematics and Science
    Study (TIMSS) 179, 188

unemployment case study 70,
    71–3
United Kingdom
  numeracy in 37–8
universe, exploring 22

videos 216
volume experiment 165

wealth 16
whole-school numeracy 151
　case study 168
　challenges to 153–5
　resources for 155–9
　strategies 160

workplace
　mathematical knowledge in 60
　numeracy at 6–10
　technology 10
World Bank 17

For Product Safety Concerns and Information please contact our EU
representative GPSR@taylorandfrancis.com
Taylor & Francis Verlag GmbH, Kaufingerstraße 24, 80331 München, Germany

www.ingramcontent.com/pod-product-compliance
Lightning Source LLC
Chambersburg PA
CBHW061438300426
44114CB00014B/1735